1st International Conference on Non-Destructive Evaluation of Composite Structures (NDECS 2017)

MATEC Web of Conference Volume 191 (2018)

Tetouan, Morocco
25 November 2017

Editor:

A. Khamlichi

ISBN: 978-1-5108-6871-7

Printed from e-media with permission by:

Curran Associates, Inc.
57 Morehouse Lane
Red Hook, NY 12571

Some format issues inherent in the e-media version may also appear in this print version.

Printed by Curran Associates, Inc. (2018)

For additional information, please contact EDP Sciences – Web of Conferences at the address below.

EDP Sciences – Web of Conferences
17, Avenue du Hoggar
Parc d'Activité de Courtabœuf
BP 112
F-91944 Les Ulis Cedex A
France

Phone: +33 (0) 1 69 18 75 75
Fax: +33 (0) 1 69 28 84 91

contact-edps@webofconferences.org

Additional copies of this publication are available from:

Curran Associates, Inc.
57 Morehouse Lane
Red Hook, NY 12571 USA
Phone: 845-758-0400
Fax: 845-758-2633
Email: curran@proceedings.com
Web: www.proceedings.com

TABLE OF CONTENTS

Optimizing the focal laws of ultrasonic arrays for non-destructive evaluation of composite panels with delamination

Mariam Achbal[1,a], Abdellatif Khamlichi[2] and Fadoua El Khannoussi[2]

[1]*Department of Physics, FS Tetouan, 93000 Tetouan, Morocco*
[2]*Department of Industrial and Civil Sciences and Technologies, ENSA Tetouan, 93000 Tetouan, Morocco*

Abstract. The use of ultrasonic arrays in the field of non-destructive evaluation of composites is a recent technique. This method offers the possibility to achieve fast inspection of components having complex geometries. As compared with the single element transducers, arrays configuration yields more effective detection of defects. However, to achieve adequate imaging by ultrasonic array elements, each element should be excited according to a specific sequence which is called the focal law. At each instant a subset of the array elements is activated and the incident waves are better directed to scattered at the defect. The reflected signals can then be more intensified. In this work, two-dimensional approximation of wave propagation problem was supposed. Array elements with linear geometry and with the elements assumed to be fully coupled with the composite panel were considered. A delamination defect was inserted between the plies of the composite panel at a given depth. The forward model enabling to simulate ultrasonic array data was built by means of the FEM. As each focal position requires a different focal law that fixes a proper protocol of excitation and reception, synthesis of these focal laws was studied. Optimization of signal focalization and reception as function of the delamination extent was achieved.

1 Introduction

Composites are made from various combinations of the three basic categories of materials as metals, polymers and ceramics to form the fourth basic category of materials. They consist of two or more separate materials combined in a structural unit. Composites are normally used because they have the required properties for a giving application that cannot be reached by any of the constituent materials acting alone. Because composites are the combination of two visibly distinct constituents into a single material, fibers and matrix, they present a large variety of damages that could be specific to one or both of the constituents or involve interaction of the two. the failures modes that often face the composites structures are matrix cracks, fiber/matrix interface failures, fiber failures, through-thickness failures (holes and cracks) and inter-ply failures such as delamination [1,2]. The composites properties allowed them to be used in a wide variety of industries, including aerospace that has demands for high-quality, cost-effective solutions in design, fabrication, and inspection. Since the aerospace structures are very sensitive to damage which can has various sources, regular inspection of the composite structures is necessary. Because of their anisotropy, the inspection of composites is not an easy task. Many of the inspection problems posed by composites can be dealt with effectively using ultrasonics [3,4]. Phased array ultrasonic testing (PAUT) is one of the advanced inspection technologies that promises to become the primary inspection technique for a large number of composites applications [5,6]. An array probe consists of a multiple transducer elements, which can be excited independently with appropriate time delays. The combination of sound waves generated by the transducer elements forms an intense sound beam. This sound beam propagates through a component and may reveal the presence of defects by reflecting the beam scattered on the flaws to a transducer. The time of exciting individual transducer elements is phased in order to direct and shape the sound beam. The variation of the pulsing timing changes condition how the sound waves combine with each other constructively or destructively.

Due to its inspection flexibility and ability to increase the inspection quality compared to the single element transducer, phased array (PA) transducers technology represents such an effective method of inspecting composite structures. This emphasizes the main motivation of this work aiming at improving the advantages that the PAUT to overcome the difficulties of inspecting composite structures.

Modeling and simulation are important tools to determine the optimal focal laws that enable to optimize the beamforming properties of the array and to evaluate the sound wave propagation through the component. The literature is rich of works about simulation models to determine the sound beam produced by PA probes. Buiochi et al. [7] developed a 3D computational method

[a] Corresponding author: mariam.achbal@gmail.com

based on the spatial impulse response and on the discrete representation concept to simulate acoustic beams generated by arrays through arbitrary shape of interfaces. They used this modeling method to optimize the beam focusing as produced by arrays and propagated through interfaces of complex geometry. This was performed by calculating the optimal focal law. Additionally, these authors compared two different focusing techniques and found that the cross-correlation technique could give more precise focalizing than the ray approximation technique.

The focal laws are one of the parameters that affect the beamforming. In the commercial PA probes that exist nowadays, the focal laws generators are based on the geometrical approach where the only parameter considered is the pitch between the single elements. The elements are approximated by a point source in the center of their aperture. Furthermore, the wave propagation process itself is modelled by geometrical acoustics. The traditional formulas to calculate the focusing delays are not affected by the width or length of the single transducer elements. The only geometrical value which influences the focal laws is the distance between the centers of adjoined elements (the pitch) [8]. In this context, Gommlich and Schubert [8] had created different simulation aperture models to study the influence of aperture active and passive size on the resulting signals in the focal point. The results had shown that the real aperture size, in particular the active size, has an important impact on the signals. These results had led to calculate improved focal laws by considering the real aperture size and the full wave phenomena in the phased array near-field.

In another work, Dheeraj et al. [9] discussed the effect of different focal law parameters including element quantity, pitch, focal depth, range of angle, and angle resolution, in order to insure an effective inspection. To do so, they analyzed C-scan plans using simulation and practical trials to derive the best scan plan in terms of detectability, resolution and sizing of defects. The study concludes that for an effective PAUT examination, focal law parameters must be validated either through practical or simulation approach. A simulation to evaluate the smallest detectable discontinuity based on effective scan plan is proposed.

The scope of this work is to optimize the focal law of an ultrasonic linear array probe in order to maximize delamination defect detection in composites. The PA ultrasonic modeling was performed by using ABAQUS Software. Different models of a composite panel made of laminated Carbon Fiber Reinforced Epoxy (CFRE) layers and containing a delamination defect at different depths were numerically analyzed. An optimization of the pitches to maximize focalization and reception of the beam will be conducted. The influence of the position of the receiver and the delamination depth will be studied.

2 Materials and methods

In the simplest cases, the elastic wave equation could have an analytical solution. But, to model wave propagation for more complicated structures and at high frequencies such as in ultrasonic waves, the analysis becomes more difficult. This is why approximate solutions as provided by numerical solutions should be searched for in that case. In structural mechanics and acoustics fields, use is often made of the finite element method as a basic modeling approach.

In this paper, modeling was carried out based on the finite element method by using ABAQUS software. The ultrasonic array probe parameters, the shape and the size of the inspected part and the materiel of the sample were defined as input data. Plain strain hypothesis was assumed and adequate boundary conditions were applied. The duration of calculation was scaled in order to avoid reflections from the boundaries.

2.1 Sample and Phased Array modelling

A 2-D panel simulation model using ABAQUS is created. This has the form of a composite plate made of seven bonded plies of CFRE laminates. A delamination defect was inserted between the plies of the panel. The flaw was modelled as a vacuum separating the materiel plies. Description of the modeled panel geometry, material characteristics, size depth and type of the defect in the sample are presented in tables 1 and 2.

Table 1. Materiel properties and panel dimensions

Materiel	Plies number	Dimensions mm^2	Structure Lay-up
Carbon Fibre Reinforced Epoxy (CFRE) Laminate	7	157.5 x 0.875	[0° ,90°, 0° ,90°,0° ,90°,0°]

Table 2. Defect characteristics

Defect	**delamination**
Defect size (mm)	17.5 x 0.4
Location/ Depth (mm)	4th ply / 0.42

In this study a linear array probe of 16 elements was modeled, see figure 1. The elements are assumed to be fully coupled with the composite panel. The excitation time signal g(t) of each element of the probe has the form of an oscillating pulse centered on the work frequency. It form is given by equation 1.

$$g(t)=e^{-at^2}\sin(2\pi ft) \qquad (1)$$

where f =1.5MHz is the central frequency, t the time and a the amplitude parameter.

The length of each element is 1.75 mm and the total length of PA patch is 28 mm.

https://doi.org/10.1051/matecconf/201819100001

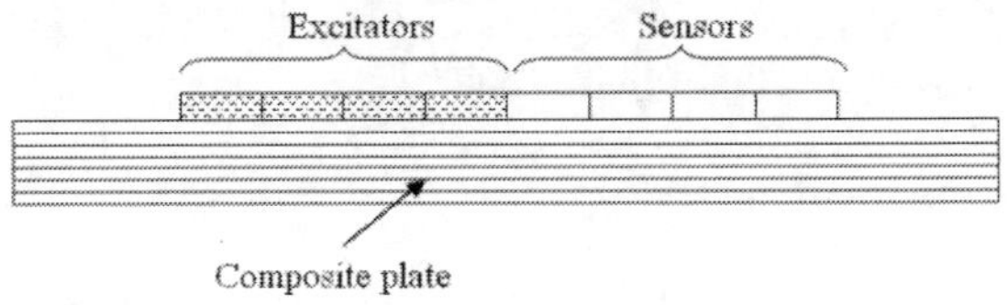

Figure 1. Testing configuration using a linear PAUT of 16 elements.

2.2 Excitation and Reception Protocol

One of the main advantages of the PA based inspection technology is its ability to perform more effectively inspection. In this work where a PA consisting of 16 linear elements is considered, the configuration of the inspection probe shown in figure is assumed to use the first 8 elements as excitators, while the response of the system is measured by the last 8 elements which play the role of sensors. Other configurations are possible by changing the number and location of excitators and sensors. In the following, only the previous pattern made of 8 excitators and 8 sensors is investigated.

To detect the presence of delamination, an indicator signal is defined as the difference between the system response in the model where the delamination defect was inserted and that one provided by a defect free model.

2.3 Additional models

To assess the effect of delamination depth two additional models with a delamination defect inserted between their plies were considered. The delamination size was kept fixed and the delamination depth is indicated in table 3.

Table 3. Defect characteristics in the additional models

Defect	Delamination	Delamination
Defect size (mm)	17.5 x 0.4	17.5 x 0.4
Location/ Depth (mm)	2nd ply / 0.42	6th ply / 0.42

3 Results and discussion

3.1 First modelling results

The PA probe and host medium systems are assumed to be linear, so they comply with the principal of superposition. The actual system's excitation input can be regarded to be the sum of each individual element excitation input function. Also, for the reason of the system's linearity, the response of the system can be calculated as the sum of individual responses of each element.

Considering the probe configuration mentioned above, the envelopes of individual responses of the system are plotted in figure 2. These correspond to the individual responses measured for each single sensor taken separately. The response consists of the reflected signal by the defect following the excitation initiated by a single element. It can be noticed that the intensity of received signals depends on the location of the excitator element. The signal intensity can be improved by adopting a PA inspection strategy through optimizing the focal laws that control the focalization and the reception of the signal.

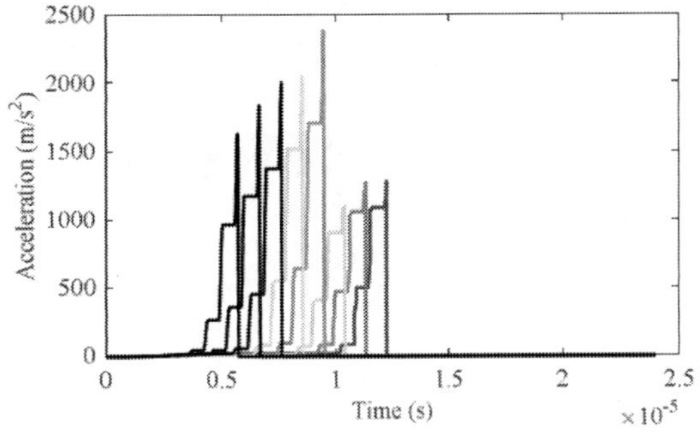

Figure 2. The envelopes of the individual responses at sensor 4 of the PA system.

3.2 Optimization results

In order to maximize the intensity of the received signal, an optimization algorithm using MATLAB was developed. The main objective of this algorithm is to calculate the optimized focal laws, and this was performed by computing the excitation phases required to receive the maximal amplitude of each individual response at the same instant.

The calculation of the optimized focal laws was achieved for each one of the eight reception sensors. The resulted signals after optimizing showed that the signal which presents the defect echo is much more intense than that associated to a single element excitation, see figure 3.

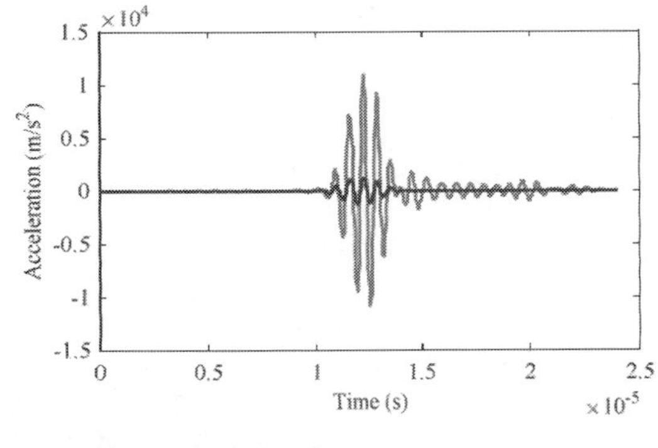

Figure 3. Received signal at sensor 4: a single exicator (blue); Phased array excitation (Red)

3.3 Effect of sensor position and defect depth

By comparing the signals received in each sensor, it could be concluded that the position of the sensor influences both the time of receiving the signal and the signal amplitude. More the position of the sensor is far from the excitation area, more the received signal amplitude is small. This is due to geometric dispersion of waves. This remark justifies why the inspection protocol takes the form of a scan as this enables the PA elements to be enough closer to the defect area.

To study the influence of the defect depth on the system's response, three different positions of the delamination defect were considered.

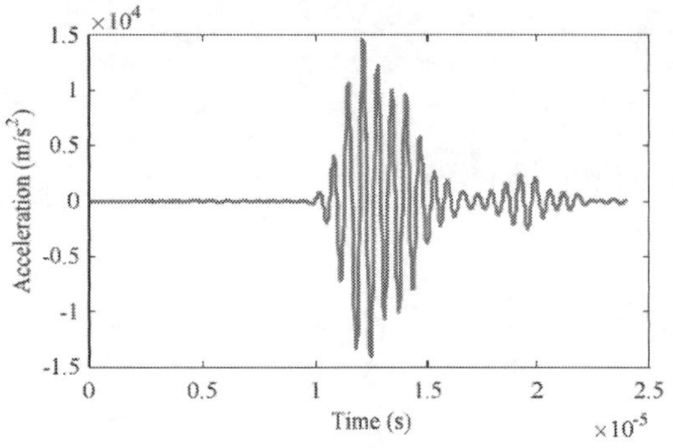

Figure 4. Received signal at sensor 4: Phased array excitation (defect in the 2nd ply)

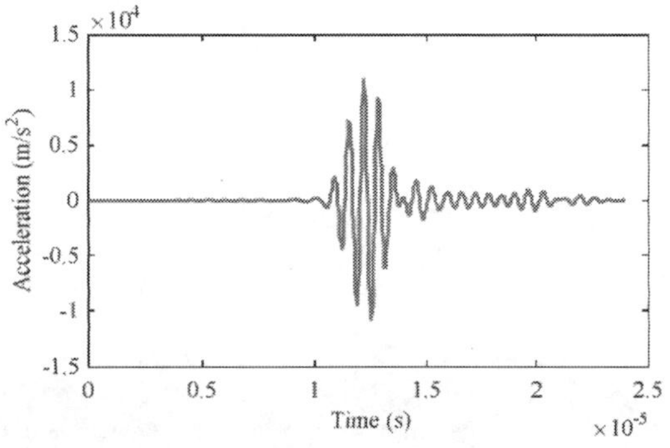

Figure 5. Received signal at sensor 4: Phased array excitation (defect in the 4th ply)

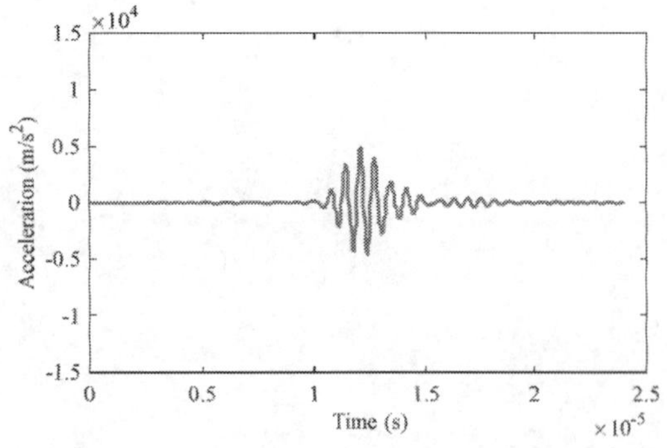

Figure 6. Received signal at sensor 4: Phased array excitation (defect in the 6th ply)

Figures 4,5,6 present successively the optimized received signals in the case where the delamination defect is in the 2nd ply, in the 4th ply and in the 6th ply of the composite panel. It can be seen that the delamination defect depth influences largely the intensity of the received signal and that the amplitude of the echo signal decreases rapidly with defect depth.

4 Conclusions

In this work, a 2-D approximation of wave propagation taking place in a composite plate was assumed in order to study detection of delamination by phased array ultrasound probes. Different models were created for the purpose of simulating this linear system. These contained a delamination defect inserted between the plies of the composite panel. A numerical solution of the problem was found using the finite element method.

The numerical simulation results obtained from the PA modelling showed that a linear array probe consisting of 16 elements with a central frequency at 1.5MHz could provide enhanced inspection of the composite panel. This could be achieved by optimizing the focal laws which control the focalization of the emitted signal and augment the amplitude of the received echo generated by reflection occurring at a defect.

The influence of the sensor position on the reflected signal received was analyzed, as well as the influence of the delamination defect depth. The results of simulations showed that the focal laws enable to intensify the echo by performing delayed firing of the excitators.

As the calculation time for the finite element simulation is rather large, it is recommended to apply in future work another numerical method in order to simulate the PA inspection process. It seems that the scaled boundary finite element method can advantageously be used for simulation of wave propagation in composites suffering from delamination.

References

1. R. F. Gibson, *Principles of composite materiel mechanics* (Taylor & Francis Group, Boca Raton, 2012)
2. C. Kassapoglou, *Modelling the effect of damage in composite structures* (John Wiley & Sons Ltd, Hoboken, 2015)
3. C. Fan, M. Caleap, M. Pan, B.W. Drinkwater, Ultrasonics, **54**, 1842 (2014)
4. A. Velichko, P.D. Wilcox, J. Acoust. Soc. Am. **127**, 2377 (2010)
5. B. Drinkwater, P. Wilcox, NDT & Int. **39**, 525 (2006)
6. O. Martínez, M. Akhnak, L.G. Ullate, F. Montero de Espinosa, NDT & Int., **36**, 57 (2003)
7. F. Buiochi, O. Martinez, L. G. Ullate, F. Montero de Espinosa, Ultrasonics **42**, 871 (2004)
8. A. Gommlich, F. Schubert. On Determination of Focal Laws for Linear Phased Array Probes as to the Active and Passive Element Size, *19th World Conference on Non-Destructive Testing,* (2016)
9. P.R. Dheeraj, I. Mohsin, S. K. Mohiuddin, S. H. Masroor, M.E. **74**, 1574 (2016)

New Generator Design with Enhanced Output Waveform for Non-Destructive Radar Sensor Measurements

Younes Ahajjam[1,3a], Otman Aghzout[1,2], José M. Catala-Civera[3], Felipe Peñaranda-Foix[3] and Abdellah Driouach[1]

[1]*Laboratory of Communications and Systems, Faculty of Sciences, Abdelmalek Essaadi University, Tetouan, Morocco*
[2]*Electronics and Microwave Group, Department of Telecommunications, ENSA, Tetouan, Morocco*
[3]*Instituto ITACA, Universitat Politécnica de València, Camino de Vera, s/n, Valencia, Spain*

Abstract. In this paper, a high power UWB monocycle generator for Non-Destructive Microwave Radar-Sensor detection is presented. The full radar system is compound with a designed UWB generator, two commercial antennas, and an oscilloscope as a receptor. The generator is composed of three essential parts-circuits, an avalanche transistor, an SRD pulse sharping and an MFN. The main idea in this work is to convert the square waveform to a driving pulse with the parameters required by the SRD pulse sharpening to obtain an output Gaussian pulse. This output pulse is converted then into a monocycle pulse. As we know, for a good range radar measurement, a high amplitude of pulse with a minimum ringing is indispensable, especially when the measurements requires many accuracies. In order to increase the pulse of the transmitter, a useful technique has been introduced; this technique consists of making the discharge path as short as possible. The applicant technique leads also to minimize the ringing in the output pulse. The measurements had been carrying out in an anechoic chamber, with the distance between 50 and 270 cm. Good Agreements between the measured and calculated results are achieved, therefore, this system can be considered very attractive in many application fields.

1 Introduction

Nowadays, there are many techniques that are used as non-destructive inspection including seismic [1-2], gravity [3-4] and electromagnetic methods [5-6]. Each technique as its advantages and drawbacks, but the electromagnetic based methods seem still to be the best ones. The reason is that because they allow for higher precision in measurement and have less dependency with regards to the environmental conditions such a temperature or humidity.

The electromagnetic method is based on the design of a specific Ultra-Wide-Band (UWB) radar sensor. The interest in UWB radar sensors systems for non-destructive application has increased significantly in recent years [7-8]. Numerous of these systems work in the time domain and possess an effective frequency spectrum of around *0.1GHz* to *10GHz* [8]. These systems present many advantages; high precision, low power consumption, and wide-bandwidth characteristics.

The principal parts in any UWB radar sensors are UWB nanosecond pulses generators. The basis of a classical UWB short pulse generator is a driver, a pulse sharpened and a pulse-forming network. On the other hand, there are various techniques to generate a nanosecond monocycle UWB pulses. A uniplanar sub-nanosecond monocycle pulse generator has been reported in [9]. A UWB monocycle pulse generator with dual resistive loaded shunt stubs was presented in [10]. In addition, an UWB monocycle pulse generator with good performance has been presented in [11]. In almost all of the previous mentioned papers, the monocycle pulse amplitude was limited, and did not enable to satisfy familiar criteria of a good radar sensor capable of providing high standard accuracy which is needed in non-destructive measurement methods.

To fulfill this task, a simple technique and useful solution have been used to improve the output pulse of radar. This technique consists in making the discharge path as short as possible in the driver parts. In this manner, a minimum of ringing around *3.6%* in overshoot has been obtained. However, and unlike the generator described by Zhong and Protiva [11-12], our pulse transmitter has the advantage of producing pulses with low ringing levels and reasonably high output amplitude. These two advantages are effective in order to satisfy completely the requirement headed for getting a system having the most accurate measurement pattern.

In this paper, an electromagnetic technique in term of non-destructive methods is presented. This method consists of a designed UWB radar sensor system having high measurements accuracy. In order to demonstrate the efficiency of the proposed design, and to validate the performance of the system, some measurements of range target are presented. The measurements are made with a distance between *50* and *270cm* in an anechoic chamber

[a] Corresponding author: youah@itaca.upv.es

in which any disturbance of parasitic signals is avoided. The pulse generator circuit is entirely fabricated on a microstrip structure, which fulfills the characteristics of being low-cost, compact and miniaturization. Demonstration that good agreement between the measured and calculated result is achieved will be given.

2 Radars System Description

There are two principal configurations in the radar systems, the bi-static and the mono-static configuration; the first one happens when the modules of transmission and reception use two separated antennas, while the second one occurs when the modules of transmission and reception form a unique antenna [13]. The mono-static presents many disadvantages and needs arrangements that are technically more difficult. Therefore, in this paper, we have chosen the bi-static configuration to deal with measurements with the purpose to improve accuracy and to get more efficient outcome. In Figure 1, the whole system having bi-static configuration is presented. In order to avoid any perturbations and interferences during the measurements, the transmitting T_x and the receiving antenna R_x have been separated by a distance of *20cm*.

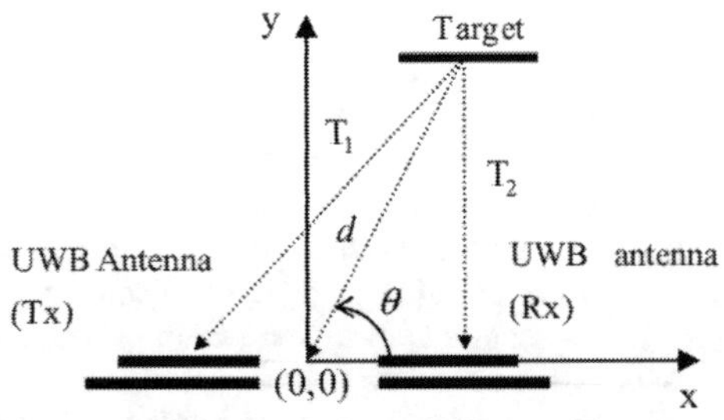

Figure 1. Bi-static system configuration of the radar.

The circuit schematic of the proposed impulse generator is shown in Figure 2. It is constructed by two principal parts: the avalanche transistor circuit and SRD pulse sharpener circuit.

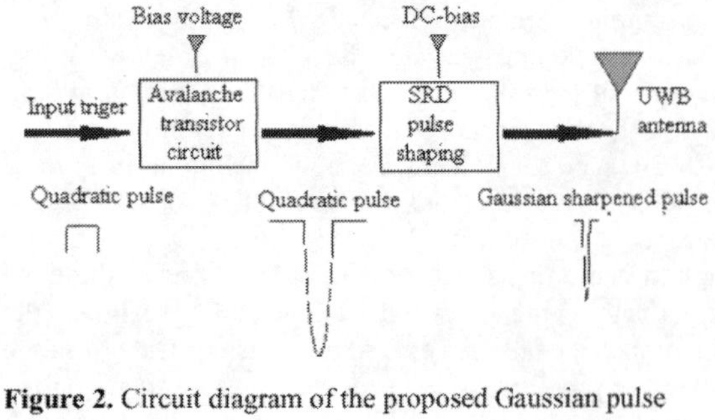

Figure 2. Circuit diagram of the proposed Gaussian pulse generator.

The first part of the generator consists of Q, C_1, R_1, C_2, and R_2 as shown in Figure 3. The details description of the generator can be found in [14].

The second part of the generator consists of an SRD sharpener and a Gaussian pulse forming the entire network as shown in Figure 4. The more detailed description can be encountered in the works presented in [14]. Furthermore, making the discharge path as short as possible, a minimum of the ringing has been obtained in the output pulse. The obtained overshoot was of the order of *3.6%*.

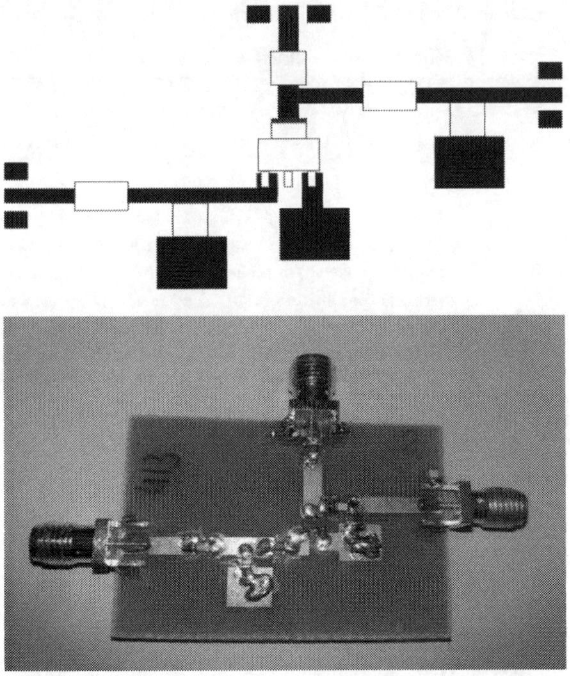

Figure 3. Layout and circuit designed for the avalanche transistor circuit.

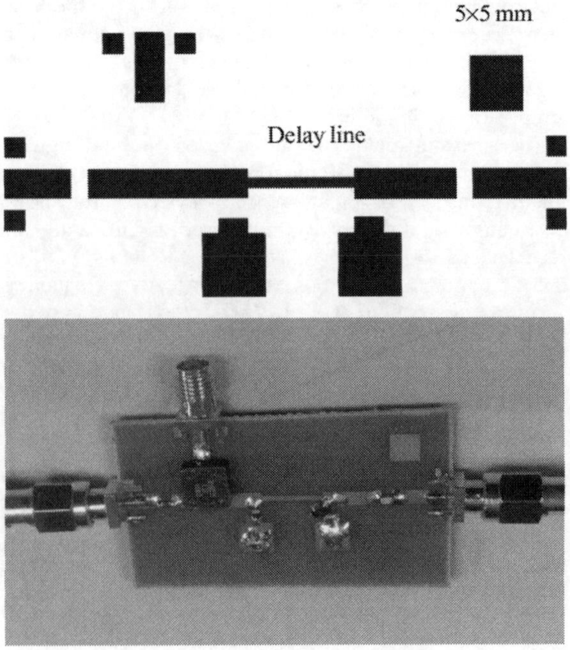

Figure 4. The designed system with SRD pulse sharpener and the corresponding equivalent Layout circuit.

3 Measurements

The developed radar sensor was constructed in bi-static configuration. The measurement setup of the sensor is shown in Figure 5. The measurements of the distance range have been made with the famous Time-of-Flight principle (ToF).

In order to avoid any parasitic signal between the transmitter (T_x) and receiver (R_x). The separation distance was elected to be *20cm*. Therefore high voltage pulses with amplitude of *6.4V* at *50Ω* are generated in all measurements.

Figure 6 shows the antennas which have been used in the measurements. They are UWB commercial antennas from RFSPACE Inc. The antennas have a high gain, low VSWR and broadband response. These antennas cover a UWB from *0.6GHz* to *6GHz*.

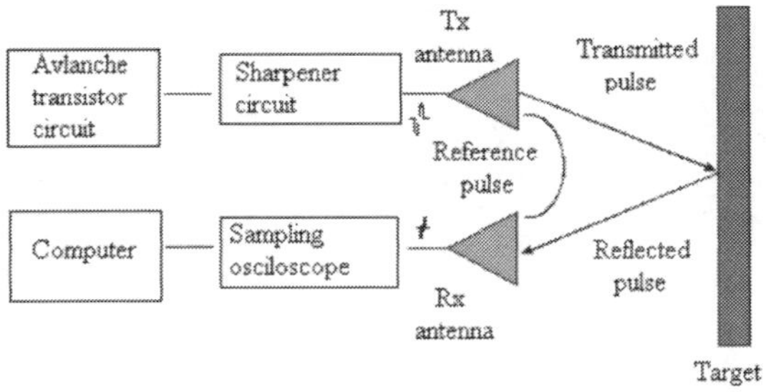

Figure 5. Measurement setup of the bi-static Radar sensor.

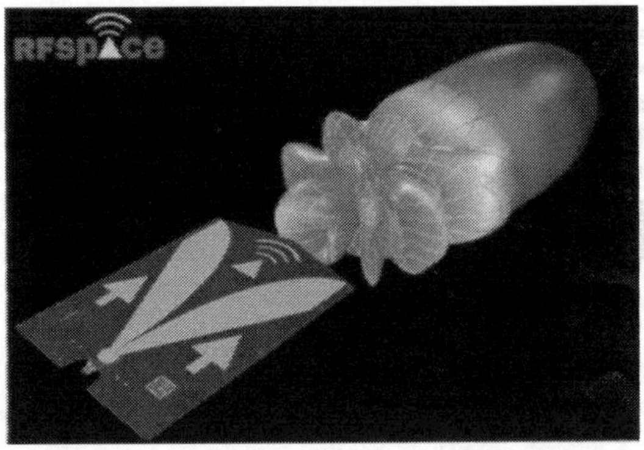

Figure 6. Photograph of the UWB antenna with the radiation pattern.

To test the efficiency of the designed radar sensor, a metal plate with a size of *30cm* x *70cm* x *3mm* has been chosen to measure the distance accuracy. For this purpose, many measurements for different distance objects ranging between *50cm* to *270cm* have been executed. The distance of the target at *50cm* has been fixed just to confirm the operability of the designed system prototype and their performance in short range application measurements. The measurements setup of the pulse radar sensor for the selected target distance and the photograph of UWB impulse generator are shown in Figure 7.

More improvement on the signal reception and better use of data at the reception can be taken out with a developed signal processing, but this was not the principal aim of this work. Our purpose in this paper was to report the hardware development and to demonstrate the workability of the radar sensor prototype in non-destructive measurements.

Figure 7. Photograph of measurement setup and a photograph of UWB impulse generator.

The time-domain response of the radar sensor measurement towards the metal plate located at *60cm* is shown in Figure 8. We observe clearly from the behavior of the curve, the presence of the reference pulse and the reflected pulse from the object; the reflected pulse was accompanied with some unwanted peaks and some ringing, this unwanted ringing occurred however after the target return is received.

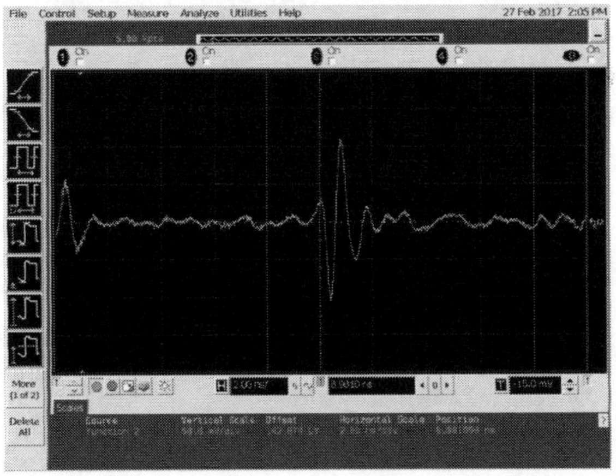

Figure 8. The time-domain response of the radar sensor measurement towards a metal plate located at 50 cm.

4 Conclusions

A UWB monocycle generator for Non-Destructive Microwave Radar-Sensor detection has been presented.

The system employs a monocycle impulse transmitter with *6.4V* in amplitude, a *0-6 GHz* sampling oscilloscope as a receiver, and two UWB antennas operating from *0.6GHz* to *6GHz*. In order to eliminate the ringing in the output pulse waveform, a useful technique has been used. This technique consists of making the discharge path as short as possible. Many measurements and test have been made to confirm the performance of this system prototype. Experimental data, obtained with this first-design system and a simple measurement procedure, also agree well with the theoretical results. Good performance together with the compactness and low cost of the impulse generator make this system very attractive as a useful tool for many non-destructive measurements.

References

1. L. Guanghui, L. Yue, Y. Baojun, IEEE Transactions on Geoscience and Remote Sensing, **55** (2017)
2. X. Dijia, B. Hongxia, J. Zheng, *25th Annual International Symposium on Personal, Indoor, and Mobile Radio Communication* (Washington DC, USA, 2014)
3. S. Fritz, K. Turkoglu, International Journal of Aerospace Engineering (2017)
4. O. B. Andersen , J. Maulik, *IEEE International Geoscience and Remote Sensing Symposium* (Quebec City, Canada, 2014)
5. Y. Yu, C.C. Chen, *IEEE International Geoscience and Remote Sensing Symposium* (Fort Worth, USA 2017)
6. Y. Fuse, M. Rostami, B. Gonzalez-Valdes, C.M. Rappaport, *11th European Conference on Antennas and Propagation* (Paris, France, 2017)
7. P. Bernardi, R. Cicchetti, S. Pisa, E. Pittella, E. Piuzzi, O. Testa, IEEE Sensors Journal, **14** (2014)
8. X. Xu, T. Xia, V. Anbu, H. Dryver, ASCE's Journal of Engineering Mechanics, **139** (2013)
9. J.S. Lee, C. Nguyen, T. Scullion, IEEE Trans Microwave Theory Tech., **49** (2001)
10. T.G. Ma, C.J. Wu, P.K. Cheng, C.F. Chou, Microwave and Optical Technology Letters **92**, 2 (2007)
11. Z. Li, X.X. Lv, Y.F. Long, T. Qin, Proc. Int. Conf. Computat. Problem-Solving (2011)
12. P. Protiva, J. Mrkvica, J. Machac, Microwave and Optical Technology Letters **52**, 11 (2010)
13. A.A.H Ameri, *Long-Range Ultra-Wideband Radar Sensor for Industrial Applications* (Kassel University, Germany, 2012)
14. Y. Ahajjam, O. Aghzout, J.M. Catala-Rivera, F. Peñaranda Foix, A. Driouach, *5th International Conference on Multimedia Computing and Systems* (Marrakech, Morocco, 2016)

Damage detection of cracks in carbon fibre reinforced composites by pulsed eddy-current testing

Abdeslam Aoukili[1,a] and Abdellatif Khamlichi[2]

[1] *Department of Physics, Faculty of Sciences at Tetouan, University Abdelmalek Essaadi, Tetouan 93030, Morocco*
[2] *Department STIC, National School of Applied Sciences at Tetouan, University Abdelmalek Essaadi, Tetouan 93030, Morocco*

Abstract. Non-destructive evaluation techniques that are based on electromagnetic methods are commonly used for inspection of metallic and carbon fibre reinforced plastics parts. Some of these inspection methods are based on eddy-currents which are generated by the electromagnetic induction phenomenon occurring under variable magnetic induction field. In eddy-currents testing, the depth of penetration into the material is controlled by the conductivity of the tested material and also the work frequency. Transient eddy-currents, also called pulsed eddy-currents, is an emerging non destructive technique that employs a pulsed excitation to induce a transient electromagnetic response from defects lying deep within a conducting structure. Such defects are difficult to inspect by conventional techniques, such as harmonic eddy-currents or ultrasonics. A large number of recent scientific publications have dealt with the theoretical understanding of the pulsed eddy-currents phenomenon and have also undertaken the design feature of appropriate probes. Finite element solution of the governing equations has been used to simulate the output signals as function of the input electrical excitation signal. Considering a B-scan strategy, simulation of a pulsed eddy-currents based probe is performed in this work with the objective to assess detectability of small defects through monitoring impedance changes of a detection probe.

1 Introduction

Composites made from Carbon Fiber Reinforced Plastics (CFRP) have superior material properties in comparison with metallic components, such as: light weight, robustness, damage tolerance, fatigue strength and corrosion resistance. CFRP based composites are being increasingly used as structural components in many industrial applications such as aerospace, shipping and automotive. However, due to their permanent exposure to external forces resulting from impact events or to the action of alternating loading, the performance of composite structures can be harmfully alterated. Defects, either resulting from manufacturing or in-service induced damage cannot be always identified by simple visual inspection. Therefore, various non-destructive testing (NDT) techniques have been used to detect and evaluate defects and damage. These include acoustic emission, ultrasonic testing, eddy current, thermography, shearography, X-ray, ultrasonic waves,...

Non-destructive evaluation (NDE) techniques based on electromagnetic methods have been commonly used for inspection of metallic parts [1]. Some of these inspection methods are based on eddy-current (EC) technique. EC are a consequence of electromagnetic induction phenomenon taking place in a conductive medium. In eddy current testing, a probe carrying varying current is placed in close proximity to the material to produce a variable electromagnetic field, and then the eddy currents are generated within the inspected materials in response to the incident electromagnetic field. These currents form closed loop circuits within the inspected material and generate in their turn magnetic fields, in the way that they oppose the effects of the applied electromagnetic field.

For damage detection in CFRP composites EC testing (ECT) has been use [2]. The authors have used rotary probes to detect fibre orientation without lateral probe movement. They have found that the proposed eddy-current scanning system is able to visualise fibre orientation, local imperfections like fibre fraction fluctuation, resin rich zones, delaminations and impact damages.

When EC are produced in a conductive material, they are not distributed uniformly throughout the inspected object. Instead, they tend to remain concentrated near to the sample surface and become less intense as the distance away from the surface increases. At a certain distance from the surface, EC effect becomes negligible and can be ignored. This phenomenon is called the skin effect which limits the depth of penetration of EC. The depth of penetration into the tested material is controlled by the conductivity of the material and also the work frequency. Transient eddy-current (TEC), also called pulsed eddy-current (PEC), is an emerging NDT technique that employs a pulsed excitation to induce a

[a] Corresponding author: aoukilia@yahoo.fr

transient electromagnetic response from defects lying deep within a conducting structure. Such defects are difficult to inspect by conventional techniques, such as harmonic eddy-currents or ultrasonics. A large number of recent scientific publications have dealt with the theoretical understanding of the TEC phenomenon and have also undertaken the design feature of appropriate probes [3-6].

To analyse response of TEC probes in terms of input and output signals, Finite Element (FE) simulation of the governing equations was considered [7].

Considering a B-scan strategy which consists in moving a probe along a straight path, the FE modelling under Comsol Multiphysics software packages is considered in the present work. A probe compound by two coils: inductive and receiver is analyzed under TEC excitation. The objective is to study delectability of a small defect through monitoring impedance changes of detection probe.

2 Materials and method

The constitutive equations governing a perfect Ohmic homogeneous and isotropic conductive medium write

$$\begin{cases} \vec{J}=\sigma\vec{E} \\ \vec{D}=\varepsilon\vec{E} \\ \vec{B}=\mu\vec{H} \end{cases} \quad (1)$$

where $\vec{E}$ is the electric field, $\vec{B}$ the magnetic induction field, $\vec{D}$ the electric displacement field, $\vec{H}$ the magnetic field, ε the permittivity, μ the permeability and σ the conductivity.

In case of a good conductive material such as a metal or CRFP, it is expedient to assume $\rho=0$ in Maxwell equations. A decoupling happens then and the following propagation-diffusion equations are readily obtained

$$\begin{cases} \Delta\vec{E}-\varepsilon\mu\dfrac{\partial^2\vec{E}}{\partial t^2}-\mu\sigma\dfrac{\partial\vec{E}}{\partial t}=\vec{0} \\ \Delta\vec{B}-\varepsilon\mu\dfrac{\partial^2\vec{B}}{\partial t^2}-\mu\sigma\dfrac{\partial\vec{B}}{\partial t}=\vec{0} \end{cases} \quad (2)$$

The magnetic permeability μ of CFRP composite material is equal to that of the free space $\mu=4\pi\times10^7\,H/m$ as the CFRP is nonmagnetic. The CFRP composites cannot be well described by the permittivity ε which is of order $10^{-11}\,F/m$, as they are conductive rather than dielectric. The conductivity is in general greater than $20\,S/m$ and can reach $2000\,S/m$. Considering low work frequencies not exceeding $10^3\,Hz$ one can verify that the propagative terms are negligible. This enables to simplify equations (2) by neglecting the propagation terms and keeping just the diffusive parts as follows

$$\begin{cases} \Delta\vec{E}-\mu\sigma\dfrac{\partial\vec{E}}{\partial t}=\vec{0} \\ \Delta\vec{B}-\mu\sigma\dfrac{\partial\vec{B}}{\partial t}=\vec{0} \end{cases} \quad (3)$$

Equations (1) and (3) yield the existence of a permanent density of currents that are circulating in the conductive material and having the form $\vec{J}=\sigma\vec{E}$, these compose the EC. For a given specimen geometry, the field $\vec{J}$ can be computed by solution of equations (3) in a given domain under specified boundary conditions.

In the presence of a singularity or a flaw in the medium, the conductivity varies and the EC distribution will show changes as compared to the pattern associated to the intact geometry. This provides a way for using EC as indicators to monitor flaw detection in CFRP made parts.

The practical implementation of EC technique for diagnosis purpose of flaws sets usually on two circuits. The first one constitutes the inductive part which is excited by a current pulse centred on a given work frequency as that shown in Figure 1. This puts variable magnetic field in the test material. The second circuit takes the form of a single coil or multiple coils sensor. Both circuits are made from metal like copper. Figure 2 depicts the EC probe proposed in this work. It consists of two separate coils. These coils are influenced by the variable magnetic field created by the EC inside the test material. The receiver coil undergoes impedance variation induced by the EC when the excitation coil is crossing a zone with variable local conductivity.

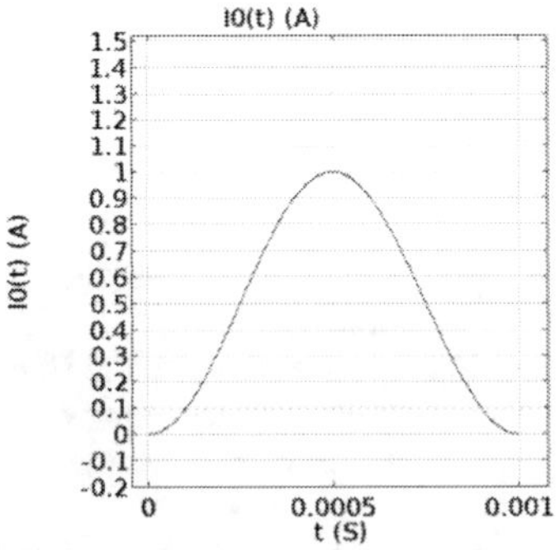

Figure 1. The transmission coil excitation function.

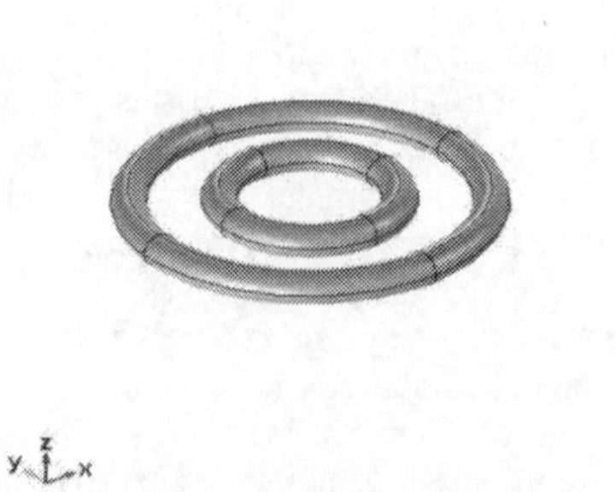

Figure 2. Geometry configuration of probe coils; the inductive in the external part and the receiver is the internal part.

Solution of equations (3) can be achieved by means of the FEM. This was performed in the present work by using Comsol Multiphysics. Here, use was made of the

https://doi.org/10.1051/matecconf/201819100003

AC/DC Module which has the capacity of simulating electromagnetic problems.

Two electrical quantities used to characterise the state of the testing coil are expressed as ohmic resistance R and inductance L. Under harmonic excitation, the reactance is related to the inductance by the following relation

$$Z_L = 2\pi f L \qquad (4)$$

with

$$L = L_s + L_m \qquad (5)$$

where f is the frequency of exciting current, while L is the combination of self inductance of the coil L_s and mutual inductance L_m .

Neglecting the capacitance, the coil impedance is given by

$$Z = \sqrt{R^2 + 4\pi^2 f^2 L^2} \qquad (6)$$

By fixing the testing frequency spectrum and lift-off of the probe, only the conductivity of the sample influences the receiver coil impedance. If an object with certain electrical conductivity is put close to this coil, the inductive magnetic field is modified which is indicated by the change of the impedance. The coil will undergo change of impedance when a defect is present in the sample. By tracking variations of the impedance, detection of defects can then be readily monitored.

3 Results and discussion

Let's consider a rectangular plate made from CFPR composite, Figure 3. The plate has the length $L = 800mm$, the width $\ell = 400mm$ and depth $h = 10mm$. The plate material characteristics are $\sigma = 95 S/m$ and $\mu_r = 1 H/m$. A defect having the form of an open parallelepiped cavity that has sides parallel to those of the plate is assumed to be present at the centre of the plate. Its dimensions are taken to be given by $L_d = 20mm$, $\ell_d = 20mm$ and $h_d = 5mm$

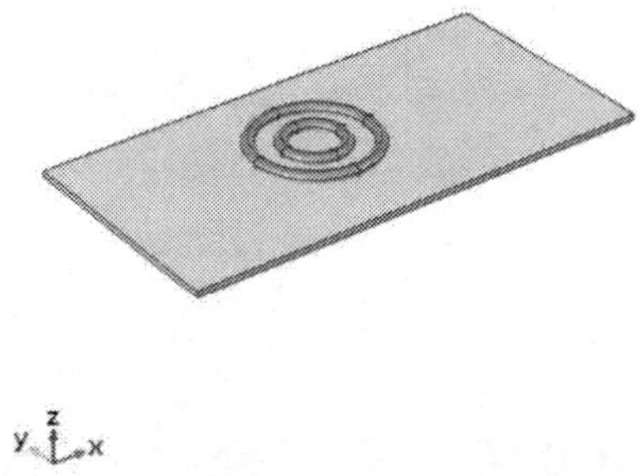

Figure 3. The test plate sample model geometry.

The inductive coil is assumed to be made from copper for which $\sigma' = 6\times10^7 S/m$ and $\mu_r' = 1H/m$. The coil base is circular with radius $5mm$. This coil is assumed to be formed by 1000 whorls having the section area $a_c = 5.7810^{-5} m^2$. It is assumed to be excited by a pulsed current of the form

$$I(t) = \frac{1}{2}\left[1 + \cos\left(2\pi\left(t - T/2\right)/T\right)\right] \text{ for } t \in [0, T] \qquad (7)$$

The sensor coil is assumed also to be made from copper. The base of receiver coil is circular with radius $2.5mm$.

A FEM model was developed for the system composed from the test plate, the probe and the ambient air. The mesh was generated automatically by using free tetrahedral elements under Comsol. The maximum size of mesh was fixed at $20mm$ for the plate and $10mm$ for the coils.

As the excitation and receiver coil are coupled to the test specimen through air according to equations (2), the environmental air which a purely propagative medium should also be meshed. Because meshing an infinite volume is not possible, it is necessary to specify a finite volume to mesh and solve for it. In the present study, the whole probe-coil setup was placed at the centre of a cube having $2000mm$ edge, see Figure 4.

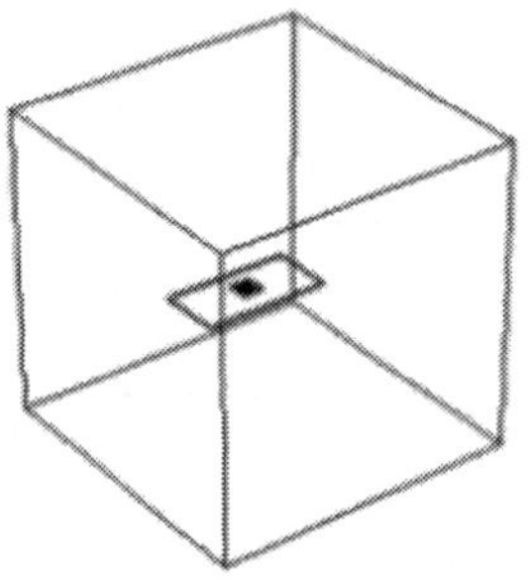

Figure 4. The complete model of the EC probe setup and test specimen.

To avoid reflection that is susceptible to take place in a finite volume domain, the default boundary condition corresponding to magnetic insulation is used at the cube boundaries. This condition forces the field to be tangential to the exterior boundaries. The air electric conductivity is fixed at $\sigma_{air} = 1S/m$. The pulse frequency is fixed at $f = 1000Hz$ which correspond to $T = 1ms$ in Eq. 7.

Both the inductive and the sensor coil are placed at the same distance from the top surface of the plate assumed to be planar and perfectly smooth. The air gap separating the coils from the inspected plate was fixed at $d = 15mm$. The EC intensity in the receiver coil was obtained by FEM post-processing.

Figures 5 and 6 give the obtained density of eddy-currents at time $t = 2T$ respectively in the defect free plate and the plate with the presence of defect, while the probe occupies the same location (over the defect) during the B-scan procedure.

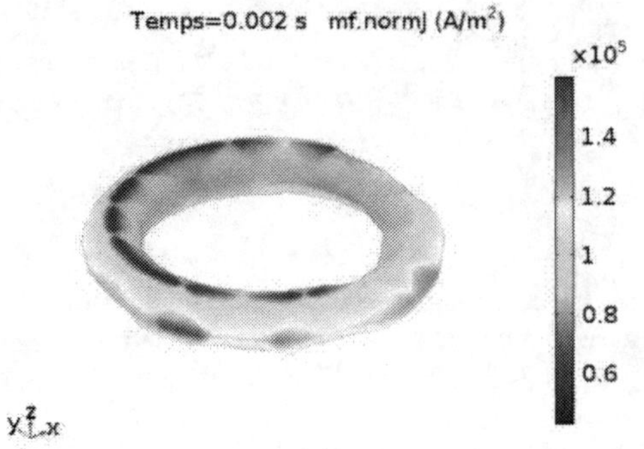

Figure 5. Current density in the receiver coil (plate without defect) at instant $t=2T$.

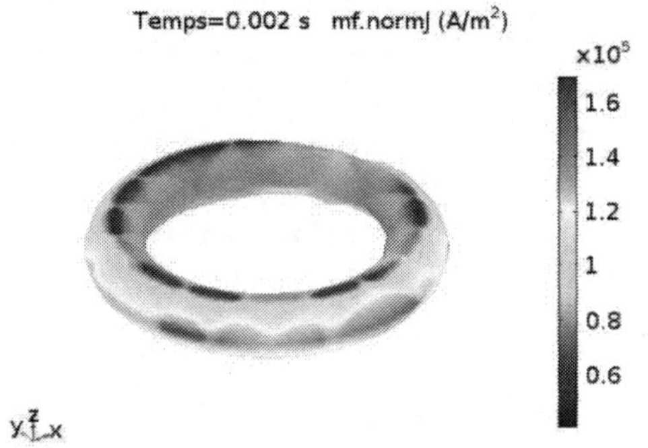

Figure 6. Current density in the receiver coil (plate with defect) at instant $t=2T$.

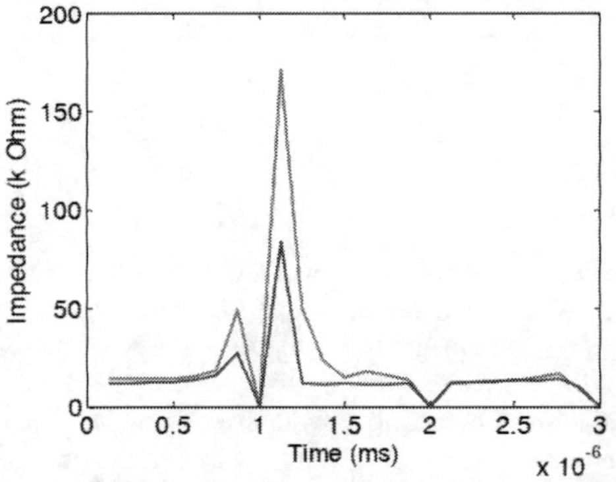

Figure 7. Comparison of impedance of receiver coil between the free defect case and the defected plate.

Figure 7 gives a comparison, in terms of impedances time evolution for the receiver coil between the case of defect free-plate and flawed plate.
Figure 7 shows that the defect induces a signature on the time variation of coil impedance that can be monitored in order to sense the defect. A defect indicator can like this be defined in order to proceed to defect detection in metallic parts. It should be noticed that while working with harmonic like excitation, this signature was very poor and cannot be readily used to detect the defect.

4 Conclusions

In this paper, a transmitter/receiver probe has been proposed to perform detection of defects appearing in composites parts made from Carbon Fiber Reinforced Plastics. Transient pulsed eddy-current based technique was implemented by using COMSOL Multiphysics software packages. Considering a B-scan inspection procedure, detection of small flaws was demonstrated in situations where conventional stationary harmonic eddy-current based technique fails to manage any detection. This was achieved through monitoring impedance variations of the receiver coil.

Further studies are needed in order to optimise detection as function of the central work frequency and probe geometric characteristics.

References

1. L. Rosado, T.G. Santos, P.M. Ramos, P. Vilac, M. Piedade, NDT&E International **51** (2013)
2. M. Gerhard, L. Rolf, K. Ole, Compos Sci Technol **61** (2001)
3. T.W. Krause, V.K. Babbar, P.R. Underhill, AIP Conference Proceedings **1581** (2014)
4. D.P.R. Desjardins, T.W. Krause, N. Gauthier, Nondestructive Testing and Evaluation International **60** (2013)
5. M. Ur Rahman, *Optimization of in-line defect detection by eddy current techniques* (PhD Thesis, Kassel University, Germany, 2010)
6. A. Rosell, G. Persson, International Journal of Fatigue **41** (2012)

Bridge deck condition assessment by using GPR: a review

Tahar Bachiri[1,a], Abdellatif Khamlichi[2] and Mohammed Bezzazi[1]

[1]*Department of Physics, Faculty of Sciences and Technology at Tangier, University Abdelmalek Essaadi, Tangier 93001, Morocco*
[2]*Department STIC, National School of Applied Sciences at Tetouan, University Abdelmalek Essaadi, Tetouan 93030, Morocco*

Abstract. Bridge deck deterioration includes delamination, corrosion of rebars, vertical cracking and concrete degradation. Ground Penetrating Radar based inspection technique was used for mapping bridge deck in order to monitor the health of this vital structure. Beginning from the 1970's, research has focused on identification of bridge deck deterioration which causes severe deficiencies by reducing seriously strength. The main purpose of this work was to review the literature existing in the field of bridge deck diagnosis by means of Ground Penetrating Radar. This enabled to survey detailed evolution of this technique, to understand the effects of key intervening parameters and to determine the kind of material disorders that this device could be used for. Bridge deck condition assessment can then be performed more successfully, based on the huge progress experienced by this particular method of inspection.

1 Introduction

Bridges are vital infrastructures for transportation activity. They require recurring maintenance and repair which can result expensive. Setting up more efficient and cost-effective methods for monitoring conditions of bridge decks is viewed to be highly imperative.

According to statistics about a quarter of constructed bridges are deficient due to stringent structural disorders or because they are functionally out of date. It was documented that the most important cause of deficiency in concrete bridge structures is corrosion of steel reinforcements [1]. The main other causes of bridge condition shortage include delamination, concrete carbonation, scaling and spalling. Delamination in concrete occurs when reinforcements corrode to the point that they expand, creating a horizontal crack.

Most corrosion processes are induced by the presence of chloride due to water and salt infiltration. Corrosion induced deteriorations need several years to rise and to become visible on the surface of the structure. The ability to perform diagnosis of bridge deck problems at their early stage can increase their service life and diminish reparation costs. Conventional inspection techniques of bridge deck consist of methods such as hammer sounding, ultrasonic pulse propagation velocity as well as infrared thermography. Recently, Ground penetrating Radar (GPR) has emerged as an outstanding technology which is both expedient and reliable for operating non-destructive bridge deck assessment. GPR is an electromagnetic investigation method which is non-destructive and fast. It can also be contact-less by using air-coupling mode. GPR can be used either in reflection or transmission mode, but reflection methods are the most frequent. GPR was recognised to be well suited for the detection of delamination and corrosion of rebars [2]. The common method used for the assessment of bridge deck condition is based on the attenuation of the reflected waves. A close relationship exists between the GPR reflection amplitudes and the chloride content in concrete and the higher the chloride content is, the lower the signal amplitudes are [3-4].

Feasibility of GPR in detecting delamination damage has received vast interest. In a comparative study conducted by Barnes and Trottier [5], GPR was found to provide better results, with up to 57% of deteriorated concrete cases detected. In another study GPR was able to identify 77% of the deteriorated areas and the depth of cracking was measured with accuracy reaching 80% [6]. Other research presented a correlation between GPR amplitudes and corrosion induced delamination [7]. It should be noticed that while delamination causes usually an attenuated GPR response, the reverse is not always true. Weak amplitudes may indicate delamination but also variations in moisture content or surface distresses. The variety of physical and environmental factors that may dampen the GPR response has been detailed in [8].

2 GPR theory and principles

In reflection mode, the principle of GPR consists of emitting an electromagnetic pulse via a transmitter antenna. This pulse propagates before undergoing reflections at the surface or interior layer boundaries of an object. The echo signal is recorded via the receiver antenna. The result of a single GPR measurement with

[a]Corresponding author: Bachiri.Tahar69@hotmail.com

the device placed at a given position is a time-series, called A-scan. This curve is characterised by amplitude and two-way-travel time of the reflected signal. The two-way-travel time or time of flight is the time required to propagate to a reflector and back to the receiver.

In an homogenous and isotropic dielectric medium, the velocity of electromagnetic waves is constant velocity and depends on the relative permittivity ε_r and the relative permeability μ_r. For materials fit for the radar method, $\mu_r = 1$ is assumed and the wave speed within a material is defined by the relative permittivity. This assumption is of course not valid for ferromagnetic materials, but using GPR for these materials is unsound. For ordinary materials the relative permittivity varies from $\varepsilon_r = 1$ (for air) to $\varepsilon_r = 84$ (for water), then electromagnetic wave velocity varies between $v = 0.11c$ (for water) and $v = c$ (for air) where c is the velocity in vacuum. The wave velocity is high, which is useful to get a high number of measurements per second.

The GPR device works by transmitting an electromagnetic energy pulse of short duration into the surface and recording reflections that occur at interfaces of materials having different dielectric properties. The reflected echoes can be monitored by focusing on reflected wave amplitude, arrival time, distortion of source wave or energy attenuation [9]. When hitting an interface, part of the energy of an electromagnetic wave is transmitted, while the other part is reflected. For a plane electromagnetic wave in a low-loss material, under vertical incidence, hitting an interface between two materials with the incident and transmitted relative primitives ε_{r1} and ε_{r2}, the reflected wave amplitude is characterized by the reflection coefficient

$$R = \frac{\sqrt{\varepsilon_{r1}} - \sqrt{\varepsilon_{r2}}}{\sqrt{\varepsilon_{r1}} + \sqrt{\varepsilon_{r2}}} \tag{1}$$

The two-way-travel time is given by

$$TOF = \frac{2d}{v} \tag{2}$$

where d is the thickness of the incident layer. Knowing the value of TOF can be used to determine the thickness, when the velocity v is identified.

The ability of the GPR method to separate single objects (lateral resolution) is limited by the size of the Fresnel region which is the area where reflected waves interfere constructively. Two points within this area cannot be distinguished by the radar signal. In general, lateral resolution increases with the work frequency f and decreases with the target depth d. The depth of wave penetration depends on the conductivity of the medium and the work frequency. Increasing the frequency of waves reduces the depth of penetration.

GPR data can be acquired and presented in various modes. Moving the antenna along a horizontal line, GPR measurements can be recorded at close intervals. The recorded time-series can be plotted side by side with the time axis pointing downwards and the horizontal axis corresponding to the line along which the antenna is moved. This display provides a B-scan.

The frequency content of the emitted and recorded radar signal is mainly defined by the antenna configuration. The choice of the appropriate antenna is important. Currently, antenna with centre frequencies between some *MHz* (low resolution and high depth of penetration) and some *GHz* (high resolution and low depth of penetration) are available.

There are several limitations restricting the use of radar and the accuracy of results. A sufficient contrast in material properties is required to enable investigation of the interface between two materials. Without such contrast the boundary will not appear in the radar raw data. This is why GPR can detect corrosion-induced delamination, but not cracking due to traffic overloading which is not visible on radargram.

3 Review of the GPR literature related to bridge deck condition assessment

The history of GPR tests in surveying bridge deck condition dates back to the 1970's [10]. In the early 1980's, GPR based bridge inspections have expanded [11-14]. The first vehicle mounted GPR system for high-ways was developed in 1985 [15]. The obtained results were really promising and the method rapidly became a routine survey tool in various rehabilitation projects of roads and bridges [16-17].

In the late 1980's and early 1990's, most of applications of GPR that were dedicated to detection of deteriorated areas in bridge decks [18-27]. These surveys were mainly conducted with high frequency 1GHz air-coupled antennas [28]. But later, ground coupled antennas were also used [29]. In the late 1990's and early 2000's, research in the field of bridge inspection focused on detection of river-bed scour around bridge piers by means of boat-mounted impulse radar survey equipment [30-32], detection of voids in post tensioned concrete bridge beams [33], mapping concrete deterioration [34-35], new antenna designed for bridge deck evaluation [37] and accuracy and reliability of GPR [5,38-39]. GPR has been well recognized as an effective and efficient technology for bridge deck inspection [40-41].

In the following, a review of the most important literature published this last decade on bridge deck deterioration assessment is performed. Maser et al. [42] considered the relationship between GPR measurements and the state of deck deterioration in order to classify stages of the deterioration process. Mercedes et al. [43] investigated the uncertainty evaluation in the measurement provided by GPR for operating frequencies of 1GHz and concluded that important differences may be related to operators. Benedetto [44] proposed a novel numerical approach for three dimensional tracking and mapping of cracks in a bridge deck. He tested the numerical approach on data acquired by using 2GHz GPR system. Alani et al. [45] conducted an experimental study on two major British bridges for identifying possible structural defects. They showed that remarkable similarity exists in the processed data concerning areas affected by ingress of moisture within the two bridges deck structures. The authors concluded that interpretation of GPR can take profit from lessons learned in earlier investigations.

Gucunski et al. [46] concentrated on bridge deck condition assessment using various non-destructive evaluation (NDE) techniques. They demonstrated the ability of NDE methods to objectively characterize deterioration progression. Dinh et al. [47] analyzed several depth correction approaches proposed to account for variation of concrete cover thickness on the rebar reflection amplitude of the GPR signal. They proposed separating normalization of the depth-dependent amplitudes into two components which was shown to significantly improve the accuracy.

Post-processing of GPR acquired data was the centre of several works. The research focused mainly on automating the process of interpretation and processing of GPR records. Dinh et al. [48] focused on how to obtain the condition map for a concrete bridge deck through the relative difference between reflection amplitudes at the top rebar layer. The authors proposed and validated in this context a new method. In another work, Dinh et al. [49] have considered attenuation of GPR signal as reflected from the top rebar mat. They presented an alternative method for performing the GPR attenuation analysis in which correlations derived from waveforms with only one reflection were employed. The method was validated and achieved better description of the overall deterioration of bridge decks.

Abouhamad et al. [50] have considered detection of corrosion based on GPR signal attenuation and developed a systematic framework of image-based analysis of bridge deck. They validated the proposed methodology and concluded that their approach can be used in a more informed decision making process through highlighting areas of actual deterioration. Shakibabarough et al. [51] have focused on the interpretation of GPR image of bridge decks and proposed a novel algorithm for detecting areas of deterioration in rebar mat signatures. The proposed algorithm was validated on many real GPR images.

More recently Dinh et al. [52] presented an automated rebar picking algorithm for GPR data of concrete bridge decks. The algorithm is based on the limited and simplified hyperbolic summation technique. The picking algorithm provided good results with accuracy beyond 98%. Dinh et al. [53] presented also an automated rebar localization and detection algorithm for performing picking of rebars based on conventional image processing techniques and deep convolutional neural networks. The process enabled to recognize and retain likely true rebar peaks and to discard likely false positive detections as the tests from twenty-six bridge decks showed an overall accuracy greater than 99%.

References

1. A.A. Sohanghpurwala, *Manual on Service Life of Corrosion-Damaged Reinforced Concrete Bridge Superstructure Elements* (Transportation Research Board, Washington, 2006)
2. N. Gucunski, A. Imani, F. Romero, S. Nazarian, D. Yuan, H. Wiggenhauser, P. Shokouhi, A. Taffe., D. Kutrubes, *The Second Strategic Highway Research Program, Nondestructive Testing to Identify Concrete Bridge Deck Deterioration* (SHRP 2 Report, Library of Congress, Washington, USA, 2013)
3. J. Hugenschmidt, R. Loser, Materials and Structures **41**, 4 (2007)
4. A. Kalogeropoulos, *Non-Destructive Determination of Chloride and Water Content in Concrete Using Ground Penetrating Radar* (PhD Thesis, Ecole Polytechnique Federale de Lausanne, Switzerland, 2012)
5. C.L. Barnes, J.F. Trottier, ASCE Journal of Infrastructure Systems **10**, 2 (2004)
6. S. Yehia, O. Abudayyeh, I. Abdel-Qader, A. Zalt, Journal of the Transportation Research Board, **2044** (2008).
7. B.M. Pailes, N. Gucunski, M.C. Brown, *Correlation of non-destructive testing results to improve assessment of corrosion and corrosion damage of a reinforced concrete deck* (In Transportation Research Board 92nd Annual Meeting, Washington, USA, 2013)
8. C.L. Barnes, J. Trottier, Journal of the Transportation Research Board, **1795** (2002)
9. S. Cardimona, B. Willeford, J. Wenzlick, N. Anderson, *Bridge Decks Condition Studies in Missouri Utilizing Ground Penetrating Radar: RDT 01-012* (Jefferson City, USA, 2001)
10. T.R. Cantor T, C. Kneeter, C (1978) *Radar and Acoustic Emission Applied to Study of Bridge Deck, Suspension Cables and Masonry Tunnel* (Transportation Research Record, Washington, USA, 1978)
11. T.R. Cantor, C.P. Kneeter, *Radars as Applied to Evaluation of Bridge Decks.* (Transportation Research Record, Washington, USA, 1982)
12. C.P.F. Ulriksen, *Application of Impulse Radar to Civil Engineering (*PhD Thesis, Lund University of Technology, Sweden, 1982)
13. G. Clemena, *Nondestructive Inspection of Overlaid Bridge Decks with Ground Penetrating Radar* (Transportation Research Record, Washington, USA, 1983)
14. D.G. Manning, F.B. Holt, *Detecting Deterioration in Asphalt Covered Bridge Decks* (Transportation Research Record, Washington, USA, 1983)
15. Morey R (1998) *Ground Penetrating Radar for Evaluating Subsurface Conditions for Transportation Facilities* (Transportation Research Board, Washington, USA, 1998)
16. T. Saarenketo, Geological Survey of Finland, **15** (1992)

17. C.R. Carter, T. Chung, T. Masliwec, D.G. Manning, Geological Survey of Canada, **90** (1992)
18. K.R. Maser, W.M.K. Roddis, Journal of Transportation Engineering **116**, 5 (1990)
19. N. Parry, J.L. Davis, Geological Survey of Finland, **16** (1992)
20. G.J. Weil, Geological Survey of Finland, **16** (1992)
21. T. Alongi, G.G. Clemena, P.D. Cady, *Condition Evaluation of Concrete Bridges Relative to Reinforcement Corrosion*, (SHRP Report, Washington, USA, 1992)
22. C.R. Carter, T. Chung, T. Masliwec, D.G. Manning, Geological Survey of Canada **90**, 4 (1992)
23. F.P. Haeni, G. Placzek, R.E. Trent, Geological Survey of Finland, **16** (1992)
24. T. Saarenketo, M.K. Söderqvist, *GPR Applications for Bridge Deck Evaluations in Finland* (Proceeding of the Non-Destructive Testing in Civil Engineering, The University of Liverpool, UK, 1993)
25. P. Maijala, T. Saarenketo, P.Valtanen, *Correlation of Some Parameters in GPR Measurement Data with Quality Properties of Pavements and Concrete Bridge Decks* (Proceedings of the Fifth International Conference on Ground Penetrating Radar, Kitchener, Canada, 1994)
26. S.G. Azevedo, J.E. Mast, S.D. Nelson, E.T. Rosenbury, H.E. Jones, T.E. McEwan, D.J. Mullenhof, R.E. Hugenberger, R.D. Stever, J.P. Warhus, M.G. Wieting, SPIE, **2946** (1996)
27. D.E. Mesher, C.B. Dawley, P.C. Pulles, *A Comprehensive Radar Hardware, Interpretation Software and Survey Methodology Paradigm for Bridges Deck Assessment* (Proceedings of the Sixth International Conference on Ground Penetrating Radar, Sendai, Japan, 1996)
28. T. Scullion, C.L. Lau, Y. Chen, *Implementation of the Texas Ground Penetrating Radar System*, (Report of TDT, Texas, USA, 1992)
29. T. Scullion, S. Servos, J. Ragsdale, T. Saarenketo, *Applications of Ground-Coupled GPR to Pavement Evaluation* (Report of TDT, Texas, USA, 1997)
30. S.G. Millard, J.H. Bungey, C. Thomas, M.N. Soutsos, M.R. Shaw, A. Patterson, NDT & E International **31**, 4 (1998)
31. M.C. Forde, D.M. McCann, M.R. Clark, K.J. Broughton, P.J. Fenning, A. Brown, NDT&E International, **32** (1999)
32. P. Inchan, L. Joengwoo, C. Woncheol, *Assessment of Bridge Scour and Riverbed Variation by a Ground Penetrating Radar*. (The 10th International Conference on Ground Penetrating Radar, Delft, The Netherlands, 2004)
33. A. Giannopoulos, P. MacIntyre, S. Rodgers, M.C. Forde, SPIE **4758** (2002).
34. N.C. Davidson, S.B. Chase, *Radar Systems for Topographic Bridge Deck Inspection* (Proceeding of the Seventh International Conference on Ground Penetrating Radar, Lawrence, USA, 1998)
35. F.A. Romero, R.L. Roberts, *Mapping Concrete Deterioration: High Speed Ground Penetrating Radar Surveys on Bridge Decks Using New Analysis Method Based on Dual Polarization Deployment of Horn Antennae* (International Bridge Conference, Pittsburgh, USA, 2002).
36. D. Huston, P. Fuhr, K. Maser, W. Weedon, Nondestructive Testing of Reinforced Concrete Bridges Using Radar Imaging Techniques (Report of The New England Transportation Consortium, New England, USA, 2002).
37. S. Cardimona, B. Willeford, J. Wenzlick, N. Anderson, *Investigation of Bridgedecks Utilizing Ground Penetrating Radar*, (ICAGT, St. Louis, USA, 2000)
38. H. Shin, D. Grivas, Transportation Research Board, **22** (2003)
39. J. Hugenschmidt, *Accuracy and Reliability of Radar Results on Bridge Decks* (Proceedings of the 10th International Conference on Ground Penetrating Radar, Delft, The Netherlands, 2004)
40. C.L. Barnes, J.F. Trottier, D. Forgeron, NDT&E International **41**, 6 (2008)
41. A. Tarussov, M. Vandry, A. De La Haza, Construction and Building Materials, **38** (2013)
42. K. Maser, N. Martino, J. Doughty, R. Birken, The Transportation Research Board, 2313, (2012)
43. S. Mercedes, H. González-Jorge, H. Lorenzo, P. Arias, Measurement, **46** (2013)
44. A. Benedetto, Journal of Applied Geophysics, **97** (2013)
45. A.M. Alani, M. Aboutalebi, G. Kilic, Journal of Applied Geophysics **97** (2013)
46. N. Gucunski, B. Pailes, J. Kim, H. Azari, K. Dinh, ASCE Journal of Infrastructure Systems (2016)
47. K. Dinh, N. Gucunski, J. Kim, T. Duong, NDT&E International, **83** (2016)
48. K. Dinh, T. Zayed, F. Romero, A. Tarussov, ASCE Journal of Bridge Engineering **20**, 6 (2015)
49. K. Dinh, N. Gucunski, J. Kim, T.H. Duong, NDT&E International, **92** (2017)
50. M. Abouhamad, T. Dawood, A. Jabri, M. Alsharqawi, T. Zayed, Automation in Construction, **80** (2017)
51. A. Shakibabarough, A. Bagchi, T. Zayad, *Automated Detection of Areas of Deterioration in GPR Images for Bridge Condition Assessment* (Fourth Conference on Smart Monitoring, Assessment and Rehabilitation of Civil Structures, Zürich, Switzerland, 2017)
52. K. Dinh, N. Gucunsk, T.H. Duong, NDT&E International, **98** (2018)
53. K. Dinh, N. Gucunski, T.H. Duong, Automation in Construction, **89** (2018)

Reliability of rail transverse flaw detection by means of an embedded ultrasonic based device

Hajar Benzeroual[1,a], Abdellatif Khamlichi[2] and Alia Zakriti[2]

[1]*Faculty of Sciences, Systems of Communication and Detection Laboratory, Department of Physics, 93000 Tetouan, Morocco*
[2]*National School of Applied Sciences, Industrial and Civil Engineering Department, 93000 Tetouan, Morocco*

Abstract. The main causes of train derailment are related to rail defects such as head surface defects, horizontal defects, vertical longitudinal split defects, star cracks at colt holes and diagonal crack in web. The most adverse form of these defects consists of transverse flaws that are either opened or internal. These flaws are developed generally in a plane which is orthogonal to the rail direction. Detection of these defects is essential in order to improve security of rail roads and to prevent catastrophic failures. Most of the inspection techniques of rails relay on eddy currents, electromagnetic induction and ultrasounds. Unlike some conventional inspection tests based on contact excitation probes that roll on the rail head, this work deals with theoretical analysis of an integrated contact-less system based on ultrasound waves. Generation of these waves was performed by using a non-ablative laser source. A rotational laser vibrometer was used to perform reception of the echoes. Through finite element modelling of the rail system under specific ultrasound excitation of the rail head, detection of transverse rail flaws was studied. Reliability of defect detection was then assessed by considering various sources of perturbations.

1 Introduction

Railways traffic is currently increasing and speed of modern trains is augmenting considerably. Maintenance of rails has become then a major challenge in order to assure sufficient safety of this mean of transportation to the occurrence of accidents. This task requires preventing rail failure which is known to yield catastrophic derailments with lots of casualties and huge economic loss [1]. Among the techniques used in the context of inspection of rails, the ultrasonic wave propagation based methods are very common [2].

Ultrasonic waves are waves having the form of very short pulses that are generated at the rail surface [3] and centred on a chosen work frequency. Usually, these frequencies range from *0.1MHz* up to *50MHz*. Ultrasonic waves propagate in the rail structure under various dispersion patterns. When crossing an internal flaw, these waves produce echoes that can be sensed to determine some indications about the defect [4-7].

Some difficulties affect however, the capability of flaw detection within the context of ultrasound based techniques, as well as by using other known techniques. This happens especially when the defect is very small. The cause is mainly due to the presence of various sources of uncertainties and the involvement of some perturbations in the studied problems. Statistics have revealed in fact that almost 75% of predictions provided by actual inspection methods were false.

In ultrasonic based inspection techniques of railways, diffraction of the incident elastic waves occurs at any defect present in the rail system. The resulting echoes can then be sensed at some reception points by means of adequate implemented sensors. Excitation can be achieved by various methods. Here, focus is on irradiation of the rail by a laser beam source, while reception of the ultrasonic wave amplitudes is performed by a laser vibrometrer [8].

To explore the best configuration for transverse rail flaws detection as based on the previous mentioned approach, simulation of ultrasound waves crossing these defects is considered in the following. Specific ultrasound excitation of the rail head is performed. The finite element method through dynamic analysis under ABAQUS software packages is used. Detection is based on change of wave signal indicator between the two rail situations: (1) the undamaged rail taken to be the reference state and (2) the rail state in the course of inspection with the presence of a localized transverse crack.

Different crack depths are considered in this study. Reliability of defect detection is assessed by considering sources of perturbations related to rail material properties and laser irradiation spot. Analysis of Variance (ANOVA) is introduced to determine the most influential factor and to analyze sensitivity of the indicator to variations of the intervening factors.

[a] Corresponding author: benzeroual.hajar@gmail.com

2 Modelling methods and reliability analysis

2.1 Modelling propagation of ultrasonic waves in rail system

A finite element model of the rail is proposed in the following. Simulation of ultrasonic elastic wave propagation in the European standard rail structure 60E1 is considered. Figure 1 shows the geometry of the rail system with a localized defect having the shape of a transverse crack.

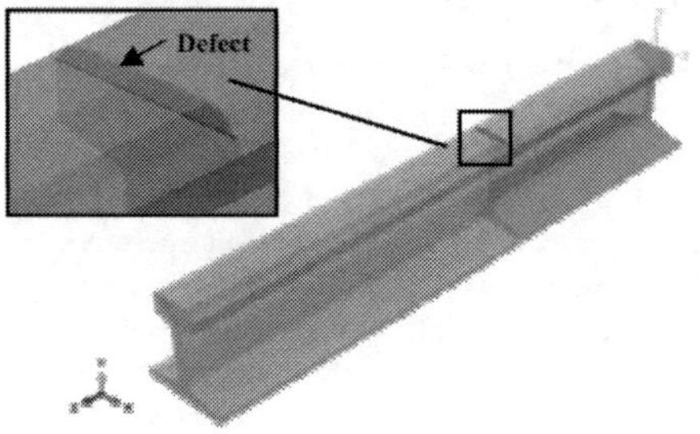

Figure 1. Schematic of a rail with localized transverse flaw.

Rail structure is assumed to be made from linear homogeneous and isotropic elastic steel material. The mechanical properties are given in Table 1. The modelled portion of the rail has length 1.3*mm*. The geometry of the defect is rectangular and has *10mm* in depth and *2mm* in width.

Table 1. Mechanical properties of the considered rail.

Young's Modulus (*GPa*)	*210*
Density (kg/m^3)	*7800*
Poisson's ratio	*0.3*
Conductivity (*W/m/K*)	*50.2*

The method of excitation is based on laser generated ultrasound. This technique is a contact-less arrangement that enables to generate ultrasonic signals remotely with the aid of lasers [9]. Short and focused laser excitation is performed on the top head of the rail to excite a small spot of the surface of the rail and to generate ultrasonic waves. In this case the spot size is chosen to be *s=0.015m*. Figure 2 shows such a configuration.

For reception of wave signals, a sensor using laser vibrometry concept is used to measure the elastic rotation occurring at any given section of the rail. This sensor is termed Rotational Laser Vibrometer (RLV). It enables to detect the reflected energy of the transmitted beam. It is also a contact-less device that allows measuring very accurately the angular velocity and angular displacement generated by elastic deformation between any pair of points on the rail structure. In this work, the RLV is calibrated to measure the normal velocity to the head surface of the rail, velocity in the y-direction (V_y in *m/s*). The sensor works with a given pair of points. Figure 3 gives the configuration of some points selected in order to measure the deformation caused by propagating ultrasonic waves.

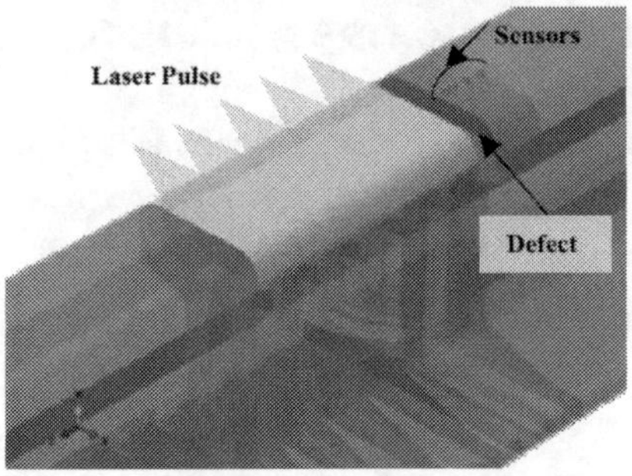

Figure 2. Position of sensors on the rail with the laser spot.

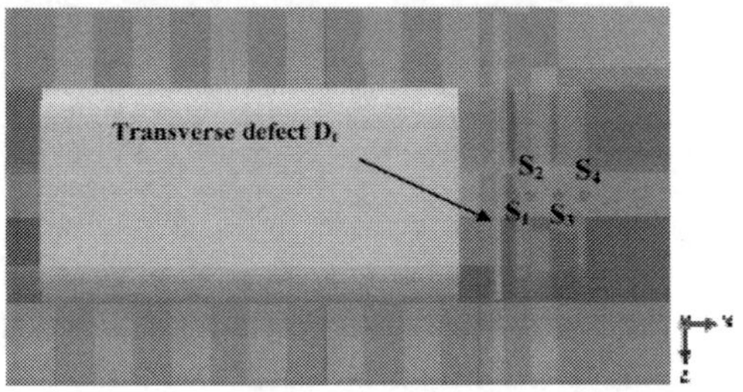

Figure 3. Position of sensors (S_i) and the defect (D_t).

2.2 Simulation of ultrasonic waves

The excitation pulse can be generated by modulation of laser spot irradiation [6]. Here, the pulse duration was fixed at $\Delta T = 450\mu s$ and a train of 20 sequential steps of the same duration was applied. The total time of computation is $100\Delta T$. The V_y amplitude obtained at each sensor location at the final instant $t_f = 45ms$ are the main information used in detection.

In order to evaluate the influence of the various parameters on the rail response, in the presence of a transverse defect, simulations were carried out according to all combinations resulting from the design levels that are indicated in Table 2. The chosen parameters include: laser spot area s, depth of the crack d and Young modulus E.

Table 2. Levels of the considered parameters.

Parameters	*s (mm)*	*d (mm)*	*E (GPa)*
Level 1	*14.85*	*9.9*	*208*
Level 2	*15*	*10*	*210*

Simulations were performed for both intact and defected rail. The considered combinations served to evaluate the relative influence of the selected factors.

Standard ANOVA procedure was used to analyze variability of the sensed wave intensity as function of variations of factors.

2.3 Reliability analysis of flaw detection

Structures are always subjected to uncertainties due to geometric variations, stochastic material properties and applied loading. These make their response to be random and circumstances under which they manifest some performance criterion are therefore not deterministic. Reliability is a pertinent tool that enables to quantify the effects of these uncertainties and to calculate the probability of failure for a given performance function. This is achieved by performing propagation of the uncertainties from the densities of probabilities of the random variables which are present at the basic level as inputs of the system to the considered performance criterion which is the actual output [10].

The performance criterion called also limit state is here related to the aptitude of detecting a transverse flaw which is present in the rail head with given characteristics. Failure of detection is considered to occur when the following condition is satisfied

$$g(X)\leq 0 \qquad (1)$$

where X designates a vector containing all the variables that are involved in the reliability analysis and g is the limit state function which separates the acceptable region with affirmative detection from that corresponding to failure of detection.

In the present case, the performance function cannot be obtained explicitly. Coupling software that predicts the actual state of the system with a reliability code is a method that can be used to perform reliability analysis. However, this process is costly [10]. There exist various methods that can be used to get a fast and robust approximation of the performance function if it is implicitly determinate. These methods enable to construct an explicit mathematical representation of the exact limit state and permit its direct evaluation as function of the basic random variables of the process.

The explicit representation model of the system is constructed by performing at first an adequate choice of trial points over the fixed domain to be explored as function of the basic variables. This is generally performed according to a design of experiment table based procedure. The simplest approach to select trial points uses a full factorial table that is generated from the selected discrete levels of the factors [11]. Using the results of simulations for these particular points, a metamodel can be derived. This provides an explicit representation of the state function over the considered domain of basic variables.

A common approach that is generally used to derive a metamodel of the system is the Response Surface Method (RSM) based method. The RSM takes usually the form of a polynomial function. Defining the densities of probabilities that describe the intervening random variables permits then to use Monte Carlo Simulation (MCS) to calculate the probability of failure.

The MCS approach enables to generate values of a random variable at the output based the probability distributions that describe variability of the inputs. By increasing the number of samples, the estimated probability of failure converges. By denoting n_f the obtained total number of failure realizations, the failure probability p_f can be estimated as

$$p_f = n_f / N \qquad (2)$$

where N is the total number of samples. The coefficient of variation of the estimate p_f is proportional to $1/\sqrt{N}$. This means that in order to achieve an estimate of p_f in the order of 0.001, approximately $N=10^5$ simulations are required.

3 Results and discussion

Considering the rail system described in section 2.1, simulations have been performed under ABAQUS for all the combinations resulting from a full factorial design of experiment table constructed on Table 2. For each combination, the amplitude of y-velocity at sensors locations (S_i, $i=1,2,3,4$), as shown in Figure 3, was obtained at the time instant $t_f = 45ms$. This was performed for both defect free rail and rail having a transverse like defect. Table 3 gives the obtained results for a rail with defect, while Table 4 gives the results for a defect free rail.

Table 3. The results of sensor 1 and 2 in (*mm/s*), for the case of rail with transverse defect.

			With transverse defect			
s(*mm*)	d(*mm*)	*E(GPa)*	S_1	S_2	S_3	S_4
15	*10*	*210*	*1.53*	*1.46*	*1.24*	*1.12*
15	*10*	*208*	*1.44*	*1.39*	*1.15*	*1.01*
15	*9.9*	*210*	*1.47*	*1.42*	*1.34*	*1.09*
15	*9.9*	*208*	*1.44*	*1.41*	*1.32*	*1.06*
14.85	*10*	*210*	*1.53*	*1.47*	*1.31*	*1.23*
14.85	*10*	*208*	*1.44*	*1.39*	*1.22*	*1.14*
14.85	*9.9*	*210*	*1.51*	*1.48*	*1.30*	*1.09*
14.85	*9.9*	*208*	*1.49*	*1.46*	*1.27*	*1.07*

Table 4. The results of sensors in (*mm/s*), for the case of defect- free rail.

		Without transverse defect			
s (*mm*)	*E (GPa)*	S_1	S_2	S_3	S_4
15	*210*	*1.98*	*1.93*	*1.72*	*1.41*
15	*208*	*1.87*	*1.84*	*1.61*	*1.27*
14.85	*210*	*2.01*	*1.99*	*1.74*	*1.46*
14.85	*208*	*1.94*	*1.91*	*1.67*	*1.32*

To perform detection, the difference between the values at two sensor locations for a rail with a defect and a defect free rail were calculated. Detection is based on variation of these amplitudes as the device crosses a defected zone. It is interesting to analyze at first how these signals vary as function of the input data: spot size, defect depth and Young modulus. Figure 4 gives the results of ANOVA performed on the obtained amplitudes as calculated from data given in Table 3.

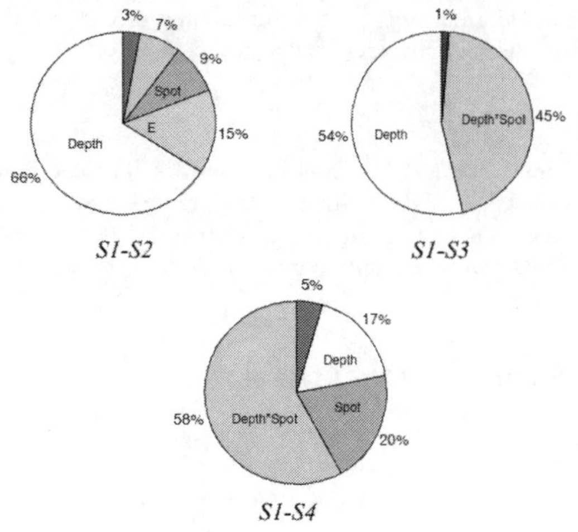

Figure 4. ANOVA results for the pair of sensors *S1-S2*, *S1-S3* and *S1-S4*; rail with a defect

Figure 4 shows that *S1-S2* configuration is sensitive to Young modulus and spot size even if it is more sensitive to defect depth than the others. The third case *S1-S4* is also sensitive to spot size. ANOVA indicates that the best configuration is provided by sensors *S1-S3* as sensitivity to crack depth is higher and effect of perturbations is reduced.

To analyze reliability of detection based on variation of the amplitudes while traversing a defect, a RSM was derived for the output associated to defect free rail and rail with a defect. The obtained regressions have all high coefficient of determination which was superior to *99%*. As an example, the obtained amplitudes for respectively rail with defect and defect free rail for sensors *S1-S3* are:

$$A_{13} = -1145.8 + 1220.2s + 1077.3d + 10.751E \\ -1152.3sd - 73.493sE + 63.25dE \tag{3}$$

$$A_{13}^{0} = -31.193 - 32.081s - 31.532E + 32.424sE \tag{4}$$

Monte-Carlo simulations were performed in Matlab for the three configurations of sensors. Figure 5 gives the obtained results corresponding to the basic variables modeled by means of the lognormal distribution for spot size and crack depth, while the Young modulus variations were distributed according to a Gaussian law. All these distributions of probabilities were centered on the median values given in Table 3 and were affected by 10% standard variation. The circles in Figure 5 indicate the maximum amplitude obtained by the considered pair of sensors while crossing a defect.

It is interesting to see that sensors configurations *S1-S3* necessitates the least resolution, while the *S1-S2* could not work without sensors of ultra-high resolution as the maximal signal amplitude represented by the red circle is too small.

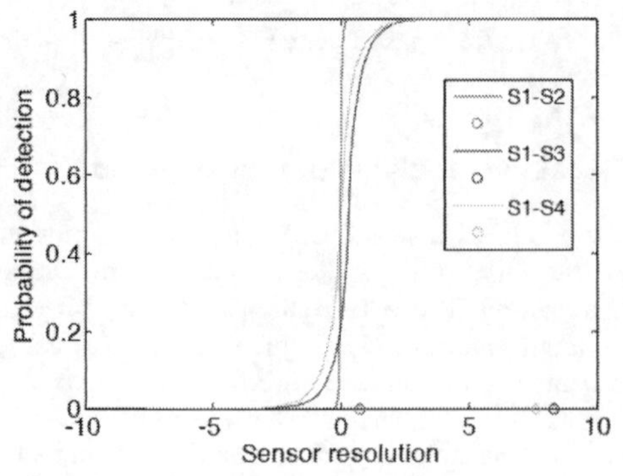

Figure 5. Probability of detection as function of the sensor resolution for the three configurations *S1-S2*, *S1-S3* and *S1-S4*

4 Conclusions

The main point revealed at this level of the study is the influence of the distance separating the sensor from the defect. Each sensor captures information depending on this distance and once the distance changes, the influence of factors (the spot and depth of the defect in particular) changes also. The obtained results have enabled to select the best configuration of sensors locations in order to operate detection of rail crack with high resolution and reduced sensitivity to uncertainties on spot size and material Young modulus.

References

1. D.F. Cannon, K.O. Edel, S.L. Grassie, K. Sawley, Fatigue and Fracture of Engineering Materials and Structures **26**, 10 (2003)
2. G. Zumpano, M. Meo, International Journal of Solids and Structures **43**, 5 (2006)
3. U. Zerbst, R. Lundén, K.O. Edel, R.A. Smith, Engineering Fracture Mechanics **76**, 17 (2009)
4. R. Clark, NDT&E International **37**, 2 (2004)
5. J.J. Marais and K.C. Mistry, Fatigue and Fracture of Engineering Materials and Structures **26**, 10 (2003)
6. S. Teidj S., A. Khamlichi, A. Driouach, E. Jacquelin, International Journal of Railway Technology **3**, 3 (2014)
7. J.L. Rose, M.J. Avioli, P. Mudge, R. Sanderson, NDT&E International **37**, 2 (2004)
8. S.J. Rothberg, M. Tirabassi, Mechanical Systems and Signal Processing **26** (2012)
9. R. J. Dewhurst and Q. Shan, Optical remote measurement of ultrasound, Meas. Sci. Technol. **10**, 11 (1999)
10. O. Ditlevsen, H.O. Madsen, Structural reliability methods, John Wiley and Sons Hoboken (1996)
11. W.J. Roux, N. Stander, R.T. Haftka, Journal for Numerical Methods in Engineering **42**, 3 (1998)

Jacketing of existing reinforced concrete structures with composites in view of seismic rehabilitation

Siham Bouras[1,a], Abdellatif Khamlichi[2] and Sabri Attajkani[2]

[1]*Department of Physics, Faculty of Sciences at Tetouan, University Abdelmalek Essaadi, Tetouan 93030, Morocco*
[2]*Department STIC, National School of Applied Sciences at Tetouan, University Abdelmalek Essaadi, Tetouan 93030, Morocco*

Abstract. Seismic rehabilitation of pre-code existing buildings requires the choice of the method of strengthening and the determination of the amount of materials to be used optimally. Accurate evaluation of the building response in terms of its capacity at the initial state and that obtained after application of some reinforcement should be performed. For regular buildings, the nonlinear static analysis procedure constitutes a powerful tool that is used to estimate seismic performance. This procedure is characterised by its high effectiveness to account for the non-linear characteristics of the materials involved and provides a direct mean to shape the capacity curve of the construction; enabling then to make the correct decision about rehabilitation task with regards to a desired performance state. In this work, the nonlinear static pushover analysis was performed by means of ZeusNL software. Use was made of the Moroccan seismic regulations RPS2000 version 2011to determine the targeted seismic demand. Considering a four floor reinforced concrete building which is undersized with regards to actual seismic regulation, jacketing with fiber reinforced composites at different reinforcement rates was analyzed. The obtained results were expressed in terms of the lateral resistance capacity and the building tip displacement. Optimal jacketing of columns was then determined.

1 Introduction

Seismic rehabilitation of existing reinforced concrete (RC) buildings can be performed by using various techniques [1]. Among these, jacketing enables to retrofit building columns by providing external confining stresses which enhance lateral performance. In particular, sheets made from Fiber Reinforced Plastic composites (FRP) can be used as external wrapping bonded on concrete members [2-4].

There are a lot of advantages that are associated to using FRP wraps: high-mechanical properties in terms of tensile strength and elasticity modulus, low unit weight, insensitivity to corrosion, easy for use as confinement jacket and shield against attack of chloride [4-5].

Modelling the constitutive behaviour of FRP confined concrete has been the subject of many investigations [6-9]. A comprehensive review of models developed to predict the axial stress-strain behaviour of FRP-confined concrete in circular sections can be found in [10]. Empirical models that have been developed to predict the strength of uniformly FRP confined concrete relate failure, for a given plain concrete strength, to the concrete strength increase due to the lateral pressure provided by the FRP. They assume generally a constant value for the confinement effectiveness coefficient. These models have been identified through using experimental results [11]. Modelling actual FRP confined concrete strength may involve complex failure criteria of concrete under triaxial stress [12]. By analyzing 20 existing models to integrate the performance of strength resulting from carbon wraps, Rousakis et al. [13] noticed many divergences.

The majority of these models focused on the monotonic axial behavior of FRP-confined concrete [14] Recently, a unified stress-strain model that enables to represent the axial behavior of circular and square/rectangular RC columns confined internally with transverse steel reinforcements (TSR) and externally with FRP or both TSR and FRP was proposed [15].

This modelling is used in this work where focus is on sizing of jacketing with the aim of realizing seismic rehabilitation of pre-code buildings through the nonlinear static pushover analysis (NSPA) approach.

Considering the Moroccan seismic regulation RPS2000 version 2011 [16], the objective is to size the FRP wrap thickness to apply to columns in order to comply with this code recommendations. A typical regular reinforced concrete building composed of four floors and which is undersized with regards to the RPS2000 seismic regulation is considered. The building response for a given FRP jacketing thickness is calculated by means of the NSPA procedure under ZeusNL software package [17]. The building seismic response is assessed in terms of the lateral resistance capacity and the roof-tip displacement. Optimisation of the FRP wrap thickness is then conducted.

[a] Corresponding author: bourassiham@hotmail.com

2 Materials and methods

2.1 Empirical modelling of uniform FRP confinement

The unified stress-strain model proposed in [15] enables to represent the axial behavior of circular and square/rectangular RC columns confined internally with TSR or externally with FRP. The confined strain ε'_{cc} and confined concrete strength f'_{cc} are derived according to

$$\varepsilon'_{cc}=\varepsilon'_c\left[1+35(I'_e)^{1.2}\right]\ ,\quad f'_{cc}=f'_c\left[1+2.4(I'_e)^{0.7}\right] \qquad (1)$$

where ε'_c is the axial strain corresponding to concrete cylinder strength, f'_c is the concrete cylinder compressive strength. The quantity I'_e is the confinement index at peak stress. It represents the equivalent confinement pressure at peak stress as normalized by the unconfined concrete strength. It is given by

$$I'_e=\begin{cases}\min\left\{\dfrac{\nu'_{cc}}{\kappa_1-\eta};\dfrac{0.15f'_c+\kappa_2\rho_{se}f_{hy}}{f'_c(\kappa_2-\eta)}\right\} & if\ \eta<\kappa_1\\ \min\left\{\dfrac{0.15f'_c+\kappa_2\rho_{se}f_{hy}}{f'_c(\kappa_2-\eta)};I'_{e,\max}\right\} & if\ \kappa_1\le\eta<\kappa_2\\ I'_{e,\max}=\dfrac{\rho_{se}f_{hy}+E_{fl}\varepsilon_{fu,a}}{f'_c} & if\ \kappa_2\le\eta\end{cases} \qquad (2)$$

where E_{fl} is the FRP lateral modulus, f_{hy} the yield strength of transverse reinforcement, $\varepsilon_{fu,a}$ the actual ultimate FRP tensile strain. In Equation (2), the quantities κ_1, κ_2, η, ν'_{cc} and ρ_{se} as given in [15] are function of f'_c, ε'_c, E_s, E_{fl}, s, s', c_x, c_y, A_{shx}, A_{shy}, ρ_{cc} and w_i, where E_s is the modulus of elasticity of transverse reinforcement, s is the TSR spacing, s' is the clear TSR spacing, A_{shx} and A_{shy} are the total cross-section areas of the transverse reinforcement perpendicular to the x and the y directions, respectively, c_x and c_y are the widths of the column's core parallel to the x and the y directions, respectively, w_i is the clear horizontal spacing between two adjacent laterally supported longitudinal bars, see Figure 1, ρ_{cc} is the ratio between the longitudinal reinforcement area and the core section area.
The FRP lateral modulus is given by

$$E_{fl}=\frac{4E_fK_ft}{c_x+c_y} \qquad (3)$$

where t is the thickness of the FRP and K_f the geometric coefficient of confinement effectiveness which is positive and given for a section with vanishing radius of the corners according to ACI 440.2R-08 [18] as function of the column's cross-section dimensions ℓ and h ($\ell\le h$), the ratio of area of longitudinal steel reinforcement to column's cross-sectional area ρ_g and the radius of the corners r_c .

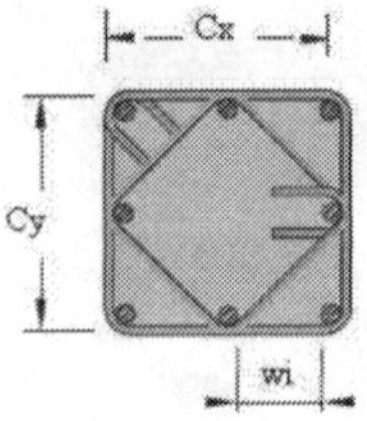

Figure 1. Core cross-section of RC member showing the longitudinal and transversal reinforcements.

In the prepeak zone $\varepsilon_c\le\varepsilon'_{cc}$, the stress-strain response of TSR and FRP-confined concrete column is given by

$$f_c=\frac{E_{ct}\varepsilon_c}{1+a\varepsilon_c+b\varepsilon_c^2} \qquad (4)$$

The postpeak branch of the stress-strain model after the point $(\varepsilon'_{cc},f'_{cc})$, for which the concrete gets into the plastic behavior phase and where $\varepsilon'_{cc}<\varepsilon_c\le\varepsilon_{cu}$, is given for $f'_{cc}<60MPa$ by

$$f_c=f'_{cc}\exp\left[k_1(\varepsilon_c-\varepsilon'_{cc})^{k_2}\right]+E_{cu}(\varepsilon_c-\varepsilon'_{cc}) \qquad (5)$$

The post FRP failure branch $\varepsilon_c>\varepsilon_{cu}$ after the point the of FRP rupture $(\varepsilon'_{cc},f'_{cc})$ is given by

$$f_c=f'_{cc,s}\exp\left[k_{1,s}(\varepsilon_c-\varepsilon'_{cc,s})^{k_{2,s}}\right] \qquad (6)$$

In equations (4) to (6), E_{ct} is the tangent elastic modulus of concrete and the quantities a, b, E_{cu}, f_{cu}, ε_{cu}, k_1, k_2 are given in [15] as function of E_{ct}, f'_{cc}, ε'_{cc}, E_{fl}, f_{hy}, ρ_{se}, ε_{fu}, $\varepsilon_{fu,a}$, h, c_x, c_y, w_i. The values of $f'_{cc,s}$, $\varepsilon'_{cc,s}$, $k_{1,s}$ and $k_{2,s}$ are derived using

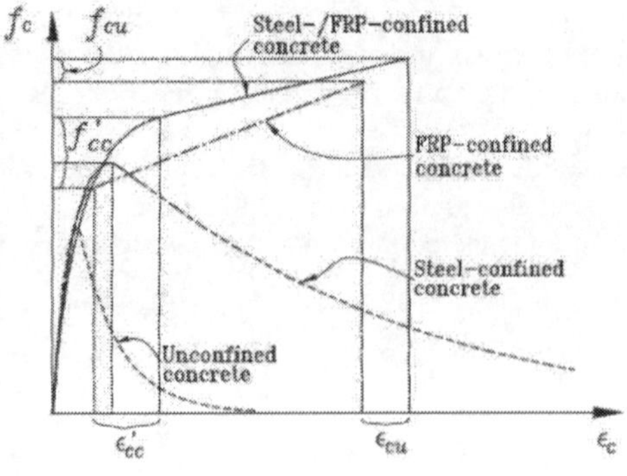

Figure 2. Stress-strain curve for TSR/FRP confined concrete.

 https://doi.org/10.1051/matecconf/201819100006

the expressions of f'_{cc}, ε'_{cc}, k_1 and k_2, respectively, for concrete confined with only transverse steel reinforcement. The module E_{cu} fixes the slope of the curve after the peak.

The stress-strain curve after the point $(\varepsilon'_{cc}, f'_{cc})$ can be of ascending or descending nature depending on the FRP lateral stiffness E_{fl}. Figure 2 shows the global stress-strain curve accounting for TSR/FRP confinements.

2.2 NSPA by means of ZeusNL software

ZeusNL was developed by Elnashai [11]. This software provides enables to run structural analyses such as conventional and adaptive pushover and nonlinear dynamic time-history. The modelling takes into account both geometric and material nonlinear behaviour. Common concrete and steel material models are available, together with a library of elements for steel and concrete composite section configurations.

In the conventional pushover analysis, the applied loads vary proportionally. The post-peak response is obtained with a displacement based control procedure. Modelling static pushover under ZeusNL software requires entering configuration of members sections, material properties and applied loadings.

As in ZeusNL the stress-strain curve given in Figure 2 is not available, an equivalent nonlinear concrete model (con2) is derived. This model is defined by the ultimate stress f'_0, the ultimate strain ε_{c0} and the confinement factor k. These are obtained from the Mander [19] model which writes

$$\frac{f_c}{f'_{c0}}=\frac{r(f'_{c0})k(f'_{c0})(\varepsilon'_{c0})^{r(f'_{c0})-1}\left[5k(f'_{c0})-4\right]^{r(f'_{c0})-1}\varepsilon_c}{(\varepsilon'_{c0})^{r(f'_{c0})}\left[r(f'_{c0})-1\right]\left[5k(f'_{c0})-4\right]^{r(f'_{c0})}+\varepsilon_c^{r(f'_{c0})}} \quad (7)$$

with $r(f'_{c0})=\dfrac{5\times10^6\varepsilon'_{c0}\left[5k(f'_{c0})-4\right]}{5\times10^6\varepsilon'_{c0}\left[5k(f'_{c0})-4\right]-k(f'_{c0})(f'_{c0})^{0.5}}$ and $k(f'_{c0})=-1.254+2.254\sqrt{1+7.94\varphi(f'_{c0})}-2\varphi(f'_{c0})$ in which $\varphi(f'_0)$ depends, for a considered direction x or y, on c_x, c_y, w_i, s, s', f_{hy}, ρ_{cc}, A_{shx} and A_{shy}.

For a given member with known TSR and FRP jacketing, the identification of the quantity f'_0 is obtained through assuming $\varepsilon_{c0}=\varepsilon'_{cc}$ and imposing that the Mander stress-strain curve be the closest possible to the actual unified stress-strain model. The optimisation is straightforward since the problem is one dimensional in terms of f'_0. The obtained confinement factor is used to scale up the stress-strain relationship throughout the entire strain range. The tensile resistance is assumed to be $f'_{t0}=6806(f'_{c0})^{0.33}$.

3 Results and discussion

The building considered is a four-story regular building that lays on a rectangular horizontal surface of $24m\times24m$. The inter story height is $3m$. The building behaviour for a given seismic direction can be represented by a four-story three-bay frame with $8m$ bay length. Figure 3 depicts the building elevation.

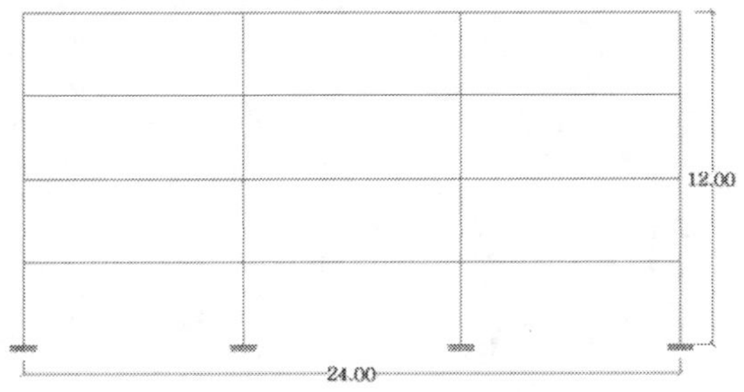

Figure 3. Vertical elevation of the four-storey RC building.

The permanent and variable loads per unit surface are respectively $G=3.3kN/m^2$ and $Q=4kN/m^2$. The active seismic gravity load is taken to be $P=G+0.2Q=4.1kN/m^2$ [16]. This RC building was designed with concrete resistance $f'_c=20MPa$ and steel characteristic resistance $f_{hy}=500MPa$. The obtained dimensions of beams are $45cm\times100cm$ with $3HA16$ compressed rebars at the top and $3HA20+3HA16$ at the bottom. For the columns, the section is $30cm\times55cm$ with $8HA20+10HA16$ reinforcement bars. The transverse rebars are $HA8$ with spacing $s=0.13m$.

The confined concrete behaviour as modelled by Eqs. (1)-(6) depends hugely on the actual FRP wrap thickness. For Carbon Fiber Reinforced Plastics (CFRP) with $E_f=70GPa$, $\varepsilon_{fu}=0.89\%$ and $\varepsilon_{fu,a}=0.85\%$, Table 1 gives as function of the CFRP thickness the identified Mander parameters. These are obtained through minimizing the difference, in the sense of L_1-norm, with respect to the unified stress-strain model [15].

Table 1. Parameters that fix the equivalence between the unified model [15] and the Mander model [19].

Thickness $t\,(mm)$	Concrete strength $f'_{c0}\,(MPa)$	Concrete strain $\varepsilon'_{c0}\,\%$	Confinement factor k
0.5	21.69	03411	1.179
1	22.91	0.3318	1.169
2	27.34	0.3053	1.139
3	37.99	0.2682	1.098

Figure 4 gives the effect of the CFRP wrap thickness, when applied uniformly to the columns, on the pushover response curve of the reinforced concrete building.

Figure 4 shows that CFRP jacketing is very efficient as it enhances lateral resistance of the building and the capacity of seismic energy dissipation. The amount of CFRP to apply depends on the seismic demand as well as on the actual state of the RC building.

https://doi.org/10.1051/matecconf/201819100006

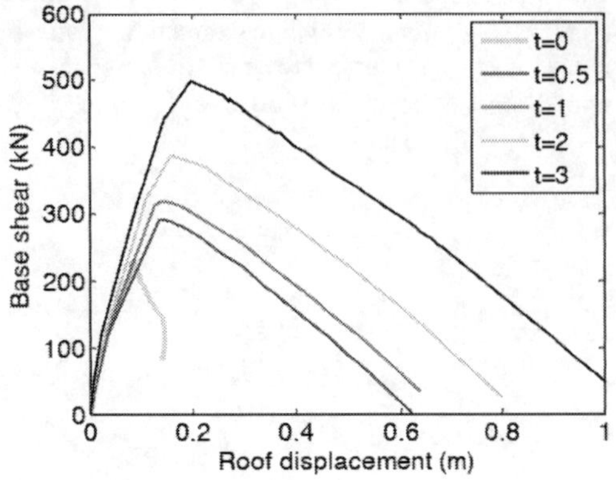

Figure 4. Pushover curve as function of the CFRP wrap thickness t in mm.

The problem is now to retrofit the pre-code building with respect to the limit states of RPS2000 which bounds the building roof displacement ratio and the maximum inter-story drift under the action of the prescribed seismic design load. The limit state associated to roof displacement threshold, δ_{roof}, should verify

$$\delta_{roof} \leq 0.004H \tag{8}$$

where H is the total height expressed in m.

For the following seismic data associated to zone 4 according to [16]: acceleration $A_{max}=1.373 m/s^2$, velocity $V_{max}=0.13 m/s$, period $T=0.4836 s$, soil foundation of class $S2$, priority class 2, ductility coefficient $K=3.5$ and damping coefficient 5%, one obtains amplification factor $D=1.1$, site coefficient $S=1$, priority factor $I=1.2$, velocity parameter $\upsilon=0.13$. So, the equivalent static lateral load is

$$F=\frac{\upsilon SDI}{K}W \tag{9}$$

where W is the total weigh of the building. The lateral loading acting on the equivalent frame is then given by $F=147.52 kN$.

Given a RFP retrofit as defined by its thickness t, the actual roof displacement δ_{roof} can be evaluated from a pushover analysis conducted under ZeusNL. Table 2 gives the obtained results

Table 2. Roof displacement as function of the FRP wrap thickness.

Thickness t (mm)	0	0.5	1	2	3
δ_{roof} (mm)	49	52	47	40	33.5

The roof displacement limit as given by equation (8) is: $\delta_{roof,\lim}=48mm$. One can see that by using RFP wrap of thickness $t=1mm$ the roof displacement is verified.

4 Conclusions

Considering a regular pre-code reinforced concrete building and the Moroccan seismic regulation RPS2000, the fiber reinforced plastic wrap thickness to apply as jacketing to columns was determined in order to fulfil the roof displacement limit requirement. This was achieved by means of the non-linear static pushover analysis under ZeusNL software package. Equivalent model that considers both transverse reinforcement and composite wrap effect on jacketed column strength was derived according to Mander model.

References

1. Y.A. Al-Salloum, Composites: Part B **38** (2007)
2. M.W. Li, ACI Struct. J. **91**, 4 (1994)
3. O. Ozcan, B. Binici, G. Ozcebe, Eng Struct **32**, 4 (2010)
4. G. Campione, Cement & Concr. Comp. **28** (2006)
5. L. Lam, J.G. Teng, C.H. Cheung, Y. Xiao, Cement & Concr. Comp. **28** (2006)
6. M. Samaan, A. Mirmiran, M. Shahawy, J. Struct. Eng. 124, **9** (1998)
7. V. Tamuzs, R. Tepfers, E. Sparnins, Mech. Compos. Mater. **42**, 2 (2006)
8. T.C. Rousakis, T.D. Rakitzis, A.I. Karabinis, J. Compos. Constr. **16**, 6 (2012)
9. H. Baji, H.R. Ronagh, C.Q. Li, Composite Structures **153** (2016)
10. T. Ozbakkaloglu, J.C. Lim, T. Vincent, Eng. Struct. **19** (2013)
11. M.F.M. Fahmy, Z. Wu, (2010), Composites, Part B **41**, 3 (2010)
12. T.C. Rousakis, A.I. Karabinis, Mater. Struct. **41**, 9 (2008)
13. T.C. Rousakis, T.D. Rakitzis, A.I. Karabinis, J. Compos. Constr. **16** (2012)
14. H.F. Isleema, D. Wang, Z. Wang, Composite Structures **185** (2018)
15. R. Eid, P. Paultre, Engineering Structures **132** (2017)
16. Secrétariat d'État de l'Habitat, *Règlement de construction parasismique RPS 2000* (Ministère de l'ATUHE, Royaume du Maroc, 2011)
17. A.S. Elnashai, V. apanikolaou, D. Lee D, *ZEUS NL - A system for inelastic analysis of structures* (Mid-America Earthquake Centre, University of Illinois, Urbana, 2003)
18. American Concrete Institute, *Guide for the design and construction of externally bonded FRP systems for strengthening concrete structures; ACI 440.2R-08* (Farmington Hills, Michigan, 2008)
19. J.B. Mander, M.J.N. Priestley, R. Park, Journal of Structural Engineering **114,** 8 (1988)

Micromechanical modelling of porous viscoelastic composite materials

Mohamed El Kouri[1,a], Abderrahmane Bakkali[1,2] and Lahcen Azrar[1,3,4]

[1]*Modeling and Control, Department of Mathematics, Faculty of Sciences and Techniques of Tangier, AEU, Tangier, Morocco*
[2]*Department of Physics, Faculty of Sciences of Tetouan, Abdelmalek Essaadi University, Tetouan, Morocco*
[3]*Research Centre STIS, team M2CS, Department of Applied Mathematics and Informatics, ENSET, MV University, Rabat, Morocco.*
[4]*Department of Mechanical Engineering, Faculty of Engineering, KAU, Jeddah, Saudi Arabia*

Abstract. Modelling and predicting the effective behavior of non-ageing viscoelastic composites have attracted the attention of many researchers. Actually, predicting the effective behavior and the macroscopic overall response while taking into account the constituents properties and shape of inclusions as well as their volume fractions is a challenging topic. In this work, porous viscoelastic composites are considered. The porous effect is introduced as a solid voided inclusions embedded in a viscoelastic matrix. The effective behavior is modelled by delayed integro-partial differential equations. The resolution of the resulting equations is done through a methodological approach based on the Volterra tensorial products and the dynamic Green's tensor. Thus, the localization equations relating the local and the global fields are derived. After that, the Mori-Tanaka mean field micromechanical model assumptions are applied to derive the Mori-Tanaka's localization tensor. Once this step is completed, the effective properties are obtained through mean field techniques. The effective properties are given through tensorial convolution products. A numerical algorithm is elaborated for the computation of direct and inverse tensorial convolution products. For the validation of the developed modelling a comparison with Laplace-Carson approach is done.

1 Introduction

Porous viscoelastic materials have been widely used due their wide range of applications: acoustic absorption, vibration isolation, low thermal conduction, electricity and magnetism shielding, as well as fire isolation. Many works have been done to model and investigate the ageing and non ageing viscoelastic materials. Two kinds of micromechanical approaches have been developed: The classic one that provides the effective behavior in the Laplace domain then a numerical inversion is performed to the time one and there is the approach that provides the effective behavior directly in the time domain.

Eshelby micromechanical models for elasticity have been extended, based on the Laplace transform, to the case of viscoelasticity by many researchers. Laws and McLaughlin [1] estimated the effective behavior of viscoelastic composites based on the Self consistent model. In the numerical results, attentions were given to the case of spherical and fibrous composites. Wang and Weng [2] extended the Mori-Tanaka model to examine the linearly viscoelastic behavior of composites with randomly oriented inclusions and of transversely isotropic ones with aligned spheroidal inclusions. Afterwards, Brinson and Lin [3], and Fisher and Brinson [4], among others, analyzed the viscoelastic behavior of viscoelastic composites using different micromechanical models. The Mori-Tanaka model was extended to the case of viscoelastic composites with interphase regions. The effect of shape of inclusions, volume fractions, and thickness of the interphase region were shown on the effective behavior. This approach based on the Laplace transform was also extended to analyze the behavior of coupled field composite materials. Jiang and Batra [5] derived closed form expressions in time and frequency domain of viscopiezoelastic composites consisting of parallel PZT cylinders of elliptic cross section embedded in a viscoelastic matrix. Azrar et al. [6] and Bakkali et al. [7] extended the Laplace transform approach to predict the time and frequency dependent behavior of viscopiezoelectric and viscomagnetoelectroelastic composite materials.

The purpose behind developing the second approach that derives the effective behavior directly in the time domain is to overcome the limitation of the first one: The case when the composite is subjected to thermomechanical loading, or one of the phases constituted the composite present an ageing behavior.

Lahellec and Suquet [8] predict the non ageing effective behavior of viscoelastic composites based on an internal variable approach. The developed modelling was reduced to a system of non linear equations at each step of time. Ricaud and Masson [9] developed a modelling based on the internal variable approach to predict ageing and non ageing viscoelastic behavior of viscoelastic composites.

The effective behavior was obtained by the resolution of a system of linear equations at each step of time. This

[a] Corresponding author: mohamed.lstm@gmail.com

work was extended by [10] to predict the ageing behavior of polycrystals. A displacement formulation was developed by Sanahuja [11] to investigate the ageing behavior of viscoelastic composite constituted of spherical inclusions.

Recently, El Kouri et al. [12] developed a new formulation based on The Dynamic Green's function techniques and on the integral equations allowing the prediction, directly in time domain, of non ageing viscoelastic composites, in the general case of anisotropic viscoelastic composites with ellipsoidal inclusions. More recently, Barthélémy et al. [13] extended the Eshelby inclusion problem for elasticity to ageing viscoelastic composites. The case of isotropic behavior with spherical inclusions and time independent Poisson ratio is considered in the numerical section.

In this work, the effective behavior and averaged fields per phase of non ageing viscoelastic composites with voided inclusions are investigated. The presented modelling gives straightforward expression of the effective behavior, in the time domain, in the general case of anisotropic viscoelastic composites with ellipsoidal voided and non voided inclusions.

This work is organized as follows: First, the governing equations are presented. Second, based on the dynamic Green's functions and on the integral equations the concentration tensors are derived. In the next section, The Mori-Tanaka mean field approach is presented and the effective properties are derived. The numerical algorithm allowing the calculation of the direct and inverse tensorial convolution products is also presented. Finally, in the numerical section, a comparison is done between the prediction obtained based on the Laplace approach and the one presented in this paper. Effective properties, global and local response are presented for different cases. The effect of volume fraction, shape of voided inclusions is shown on the behavior of the considered viscoelastic composites.

2 Governing equations

In this work, voided viscoelastic composites constituted of voided inclusions embedded in a viscoelastic matrix are considered. A representative volume element (REV) Ω is considered. At the boundary $\partial\Omega$, the macroscopic strain evolution $E(t)$ is linked with the displacement $u(x,t)$ by the following relationship $u(x,t)=E(t).x$. The non ageing behavior at each $x\in\Omega$ is modelled by a Stieltjes integral relating the stress tensor evolution $\sigma(x,t)$ to the strain tensor history $\varepsilon(x,t')$ as follows

$$\sigma_{ij}(x,t)=\int_0^t C_{ijkl}(x,t-t')\frac{\partial\varepsilon_{kl}(x,t')}{\partial t'}dt' \quad \forall x\in\Omega \qquad (1)$$

where $C_{ijkl}(x,t-t')$ is the relaxation tensor.

A compact form of Equation (1) is given by the following convolution product in the Stieltjes space

$$\sigma_{ij}(x,t)=\left[C_{ijkl}(x)*d\varepsilon_{kl}(x)\right](t) \quad \forall x\in\Omega\,\forall t>0 \qquad (2)$$

The compatibility equations, when the small displacement are considered, are given by the following expression

$$\varepsilon_{ij}(x,t)=\frac{1}{2}\left[u_{i,j}(x,t)+u_{j,i}(x,t)\right] \quad \forall x\in\Omega \qquad (3)$$

where $u_i(x,t)$ is the displacement components. In the absence of the body force, one can write the equilibrium equations as follows

$$div\left[\sigma(x,t)\right]=0 \quad \forall x\in\Omega \qquad (4)$$

Considering the symmetry of the tensors σ,ε,C and based on the previous equations, one can derive the following system of partial differential equation

$$\left[C_{ijkl}(x)*du_{k,l}(x)\right]_{,j}(t)=0 \quad \forall x\in\Omega \qquad (5)$$

3 Integral equations and localization

Let consider a homogeneous fictitious media which has the relaxation moduli $C^0(t)$. The local relaxation moduli is given by the following expression

$$C_{ijkl}(x,t)=C^0_{ijkl}(t)+\delta C_{ijkl}(x,t) \qquad (6)$$

in which x is the position vector in the considered media and δC is the deviation part.

Using the above perturbation equation, based on the dynamic Green's tensor and after some mathematical development the following localization equation is obtained [12]

$$\varepsilon^0_{mn}(t)=\left[\left(I_{mnkl}H(t)-\frac{1}{\Omega_I}\left[T_{mnij}*dC^0_{ijkl}\right](t)\right)*d\varepsilon^I_{kl}\right](t) \qquad (7)$$

with $T_{mnij}(t)$ is the dynamic viscoelastic interaction tensor given by:

$$T_{mnij}(t)\int_{\Omega_I}\int_{\Omega_I}\Gamma_{mnij}(x-x',t)d\Omega' d\Omega \qquad (8)$$

In this paper, the Mori-Tanaka model, presented in the next section, is considered. The derived concentration tensor in this section is the main step of the micromechanical modelling. The difficulty behind the calculation of the concentration tensor is that a direct and an inverse tensorial convolution product has to be computed. For this reason, a numerical algorithm is developed.

4 Effective properties and numerical algorithm

The Mori-Tanaka mean field approach is one of the most used micromechanical models. It is known by its accurate predictions, easy to implement and gives explicit expression of the effective properties. The expression of the time dependent Mori-Tanaka localization tensor is obtained by replacing $\varepsilon^0(t)$ and $C^0(t)$ of the reference media in equation (7) by the evolution strain $\varepsilon^M(t)$ and relaxation moduli $C^M(t)$ of the matrix. Therefore, one can write the following relationship relating the local evolution strain in the voided inclusion with the evolution strain in the matrix through the so called the dynamic localization tensor

$$\varepsilon^I_{kl}(t)=\left[A^{I(MT)}_{klmn}*dE_{mn}\right](t) \quad (9)$$

where $A^{I(MT)}_{klmn}$ is the Mori-Tanaka dynamic localization tensor and its expression is given by

$$A^{I(MT)}_{klmn}(t)=\left(H(t)I_{mnkl}-\frac{f_M}{\Omega_I}\left[T_{mnij}*dC^M_{ijkl}\right](t)\right)^{-1} \quad (12)$$

where f_M is the volume fraction of the matrix and Ω_I is the volume of the voided inclusion.

Now that the localization tensor is derived, average techniques are used to obtain the effective properties and overall behavior.

The overall behavior of the composite is modelled by

$$\Sigma_{ij}(t)=\left[C^{eff}_{ijkl}*dE_{kl}\right] \quad (11)$$

Based on average techniques [12], the following expression of the effective properties is obtained

$$C^{eff}_{ijmn}(t)=C^M_{ijmn}(t)-f^I\left[C^M_{ijkl}*dA^{I(MT)}_{klmn}\right](t) \quad (12)$$

The average evolution strain in the matrix is obtained through the following expression

$$\varepsilon^M_{kl}(t)=\left[A^{M(MT)}_{klmn}*dE_{mn}\right](t) \quad (13)$$

$A^{M(MT)}_{klmn}$ is obtained from the equation that verifies the localization tensors

$$f^M A^{M(MT)}_{klmn}+f^I A^{I(MT)}_{klmn}=H(t)I_{klmn} \quad (14)$$

Based on these equations, one can predict the overall and local behavior of the considered viscoelastic composite for different shape of inclusions. Various laws modelling the behavior of the viscoelastic matrix could be considered.

The main difficulty of this work is the calculation of the direct and inverse tensorial product. The algorithm is based on the discretization of time through on a trapezoidal algorithm [12]. Based on this trapezoidal algorithm, one can transform the convolution product in the Stieltjes space to a matrix-vector product as follows

$$\left[\Sigma_{ij}\right]_{(n+1)}=\left[C^{eff}_{ijkl}\right]_{(n+1)\times(n+1)}\left[E_{kl}\right]_{(n+1)} \quad (15)$$

where $\left[\Sigma_{ij}\right]_{(n+1)}$ and $\left[E_{kl}\right]_{(n+1)}$ are vectors of $(n+1)$ elements and $\left[C^{eff}_{ijkl}\right]_{(n+1)\times(n+1)}$ is a $(n+1)\times(n+1)$ matrix.

More details about this algorithm are given in [12]. The presented numerical algorithm allows one to compute the Mori-Tanaka estimate of two phase viscoelastic composites in the general case of ellipsoidal inclusion and anisotropic behavior.

5 Numerical results

In this section, the numerical results of viscoelastic composites consisting of voided inclusions embedded in a non ageing viscoelastic matrix are presented. The viscoelastic matrix is considered to be isotropic. The bulk K and the shear G moduli of the viscoelastic matrix are described by the Prony series.

$$G(t)=G_\infty+\sum_{j=1}^{N}G_j\,exp\left(-\frac{t}{\xi_j}\right) \quad (16)$$

where G_∞ is the rubbery asymptotic modulus. G_j and ξ_j are the relaxation moduli and the relaxation time. The used Prony series coefficients are given in [12].

For the sake of comparison, the results predicted based on the Laplace transform approach and on the presented modelling are compared. More details about the Laplace transform approach could be found in [12]. Based on this approach and the developed modelling for a non ageing linear viscoelastic composite consisting of spherical voided inclusions embedded in a viscoelastic matrix, Figure 1 gives the predicted results for the volume fraction of void $f^I=0.3$. One can see that the two approaches predict the same results.

Based on the developed modelling for a viscoelastic composite consisting of voided inclusions embedded in a viscoelastic matrix and for the volume fraction of void fixed at $f^I=0.2$, Figure 2 gives the effective moduli as function of the different shapes of voided inclusions. The effect of the later is shown on the behavior of the considered composites.

For the particular case of spherical voided inclusions, Figure3 illustrates the effect of the void volume fraction on the effective moduli.

Now to show the global and local response of the considered viscoelastic composites, the following macroscopic uniaxial strain is applied

$$E_{11}(t)=\begin{cases}10^{-4}t; & t<5s\\ 5\times10^{-4}; & t\geq5s\end{cases} \quad (17)$$

In Figure 4, a viscoelastic composites constituted of spherical voided inclusions embedded in a viscoelastic matrix is considered. The global and local response of the composites when it is subjected to the uniaxial strain is presented for different void volume fractions.

https://doi.org/10.1051/matecconf/201819100007

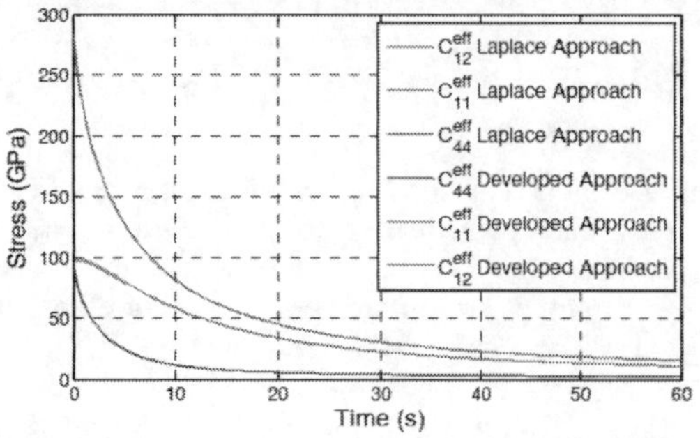

Figure 1. Effective moduli C_{12}, C_{11} and C_{44} as predicted by Laplace transform and non ageing linear viscoelastic model.

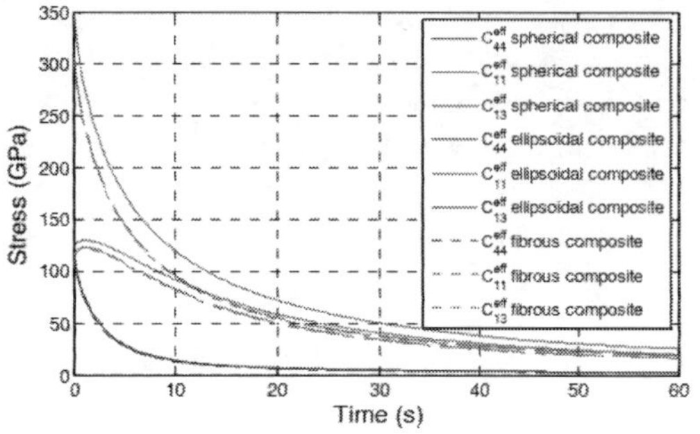

Figure 2. Effective moduli C_{44}, C_{11} and C_{13} as function of the shape of voided inclusions.

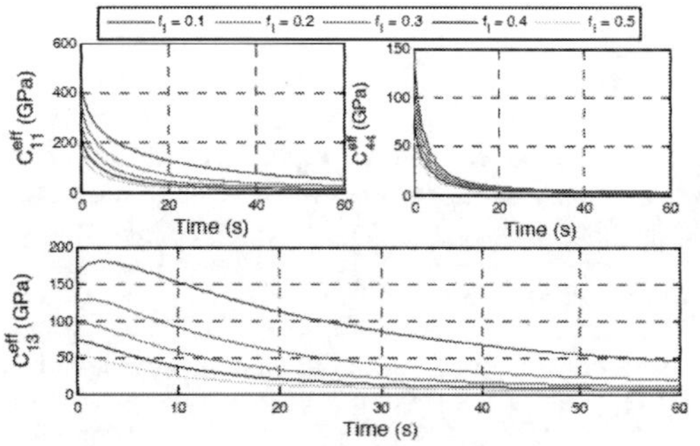

Figure 3. Effective moduli C_{11}, C_{44} and C_{13} as function of the void fraction for spherical voided inclusions.

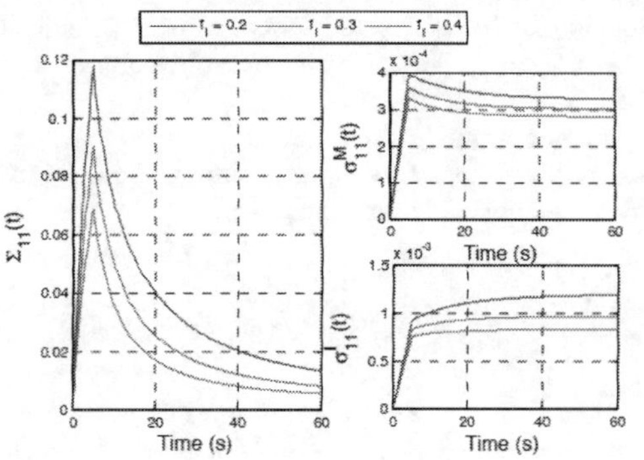

Figure 4. The global and local response of a viscoelastic composite, consisting of voided inclusions embedded in a viscoelastic matrix, subjected to uniaxial strain.

6 Conclusions

In this paper a micromechanical modelling based on the dynamic Green's functions and the Mori-Tanaka mean field approach is presented to analyze the effective behavior of viscoelastic composites constituted of voided inclusions.

The presented modelling permits a straightforward derivation of the effective behavior in the time domain. The difficulty of this work lies in the fact that a direct and inverse convolution tensorial product has to be calculated.

A numerical scheme is presented for this purpose. In the numerical section, a viscoelastic composite consisting of voided inclusions embedded in a viscoelastic matrix is considered. A comparison between the developed modelling and Laplace transform approach is presented. Different effective moduli are presented and the effect of the volume fraction and the shape of the voided inclusions is shown on the them. The global and local response is also investigated when the considered composite is subjected to a uniaxial strain.

References

1. N. Laws, R. McLaughlin, Proceedings of the Royal Society of London A: Mathematical, Physical and Engineering Sciences **359**, 251 (1978)
2. Y.M. Wang, G.J. Weng, Transactions of the ASME **59**, 510 (1992)
3. L. Brinson, W. Lin, Composite Structures **41**, 353 (1998)
4. F.T. Fisher, L.C. Brinson, Composites Science and Technology **61**, 731 (2001)
5. B. Jiang, R. Batra, Journal of intelligent material systems and structures **12**, 847 (2001)
6. L. Azrar, A. Bakkali, A.A. Aljinaidi, Composite Structures **113**, 281 (2014)
7. A. Bakkali, L. Azrar, A.A. Aljinaidi, Journal of Intelligent Material Systems and Structures (2016)
8. N. Lahellec, P. Suquet, International Journal of Solids and Structures **44**, 507 (2007)
9. J.M. Ricaud, R. Masson, International Journal of Solids and Structures **46**, 1599 (2009)
10. R. Masson, R. Brenner, O. Castelnau, Comptes Rendus Mécanique **340**, 378 (2012)
11. J. Sanahuja, International Journal of Solids and Structures **50**, 2846 (2013)
12. M. El Kouri, A. Bakkali, L. Azrar, Applied Mathematical Modelling **40**, 4302 (2016)
13. J.F. Barthélémy, A. Giraud, F. Lavergne, J. Sanahuja, International Journal of Solids and Structures **97-98**, 530 (2016)

Effects of plies orientations and initial geometric imperfections on buckling strength of a composite stiffened panel

Ikram Feddal[1, a], Abdellatif Khamlichi[2] and Koutaiba Ameziane[1]

[1]*Physics of Condensed Matter Laboratory, Faculty of Sciences, Tetouan 93002, Morocco*
[2]*Systems of Communication and Detection Laboratory, ENSA Tetouan, 93030 Tetouan, Morocco*

Abstract. The use of composite stiffened panels is common in several activities such as aerospace, marine and civil engineering. The biggest advantage of the composite materials is their high specific strength and stiffness ratios, coupled with weight reduction compared to conventional materials. However, any structural system may reach its limit and buckle under extreme circumstances by a progressive local failure of components. Moreover, stiffened panels are usually assembled from elementary parts. This affects the geometric as well as the material properties resulting in a considerable sensitivity to buckling phenomenon. In this work, the buckling behavior of a composite stiffened panel made from carbon Epoxy Prepregs is studied by using the finite element analysis under Abaqus software package. Different plies orientations sets were considered. The initial distributed geometric imperfections were modeled by means of the first Euler buckling mode. The nonlinear Riks method of analysis provided by Abaqus was applied. This method enables to predict more consistently unstable geometrically nonlinear induced collapse of a structure by detecting potential limit points during the loading history. It was found that plies orientations of the composite and the presence of geometric imperfections have huge influence on the strength resistance.

1 Introduction

In various fields of engineering, composite stiffened panels are used mainly due to their high strength-to-weight and stiffness-to-weight ratios. They have also good impact energy and sound absorption, and often low production cost. However, every structure is prone to stability problems and buckling before it reaches the geometrically linear strength limit. The global or the local buckling phenomena may occur and lead to the collapse of the structure well before that ideal limit load. In general, buckling happens when slender structures such plates or beams are subject to specific loading patterns such as an axial load applied in compression. Stiffened panels made by assembling columns and shells constitute an optimized configuration which is intended to improve resistance to buckling. For a secure design, it is important to understand the buckling behaviour and to relate it to geometric and material properties as well as the various imperfections affecting the structure. Geometric imperfections include initial imperfections related to the fabrication process. They can be distributed or localized such as initial distortions resulting from assembling process or impact happening during service life.

From the mechanical point of view, buckling of composite stiffened panels is very intricate and involves complex interactions. Various methods have been proposed to tackle this issue. They are based on experimental, numerical and computational approaches. The critical load of laminated composite plates as calculated by a linear buckling analysis was recognized to under-estimates the load-carrying capability of the system. Laminated composite plates possess load-carrying capacity after buckling which is generally several times that of the critical buckling load. Thus, the real practical limits of the load-carrying capability can be determined only by a nonlinear step-by-step analysis where the post-buckling behaviour is considered. This is termed also the strength resistance analysis of the structure.

Considerable efforts have been made for the numerical analysis of the buckling and post-buckling analysis over the years. Many studies have been dedicated to investigate the buckling related phenomena in the case of composite structures. Leissa [1] gave a review of the buckling and postbuckling studies of composite laminated plates up to 1986, and then he reviewed the development of buckling analysis of laminated composite plates with linear effective constitutive properties. Zhang and Yang [2] gave a review of the developments achieved for laminated composite plates between 1990 and 2009 and based on the finite element analysis. This literature review included buckling and post-buckling analysis of composite laminated plates. The buckling and post-buckling behaviour of unidirectional and cross-ply composite laminated plates with multiple delaminations has been studied in [3].

[a] Corresponding author: ikramfeddal@gmail.com

Non-linear finite elements analyses have been performed by using an incremental-iterative method. Large displacements and contact constraints between sub-laminates have been taken into account. The authors found that delamination length and position as well as stacking sequence of the plies influence largely the critical load of the plate. More recently, buckling and post-buckling of thick composite plates having large initial geometrical imperfections have been studied in [4]. The authors have used the theory of large transformations based on the second Piola-Kirchhoff stress and Green-Lagrange strain and considered plates exhibiting large initial imperfections. Liang and Sun [5] used a novel reduced-order modelling strategy based on the Koiter-Newton method to analyse buckling and post-buckling of laminated composite plates with delamination. Various laminated plates with different thicknesses, delamination lengths and stacking sequences were considered and enabled to validate the good performance of the method in terms of numerical accuracy and performance. Namdar and Darendeliler [6] investigated buckling, post-buckling and progressive failure of composite laminated plates by using both numerical and experimental methods. Buckling load, load-displacement relations for post buckling and maximum out-of-plane displacements of the plates were determined. The numerical results were compared with experimental findings for different laminates. The buckling resistance was found to depend largely on the orientation of plies with regards to the compressive load. A good agreement between numerical and experimental results was obtained for buckling load and post-buckling behaviour especially for the laminates with uni-directional tapes. Heidari-rarani and Khalkhali-sharifi [7] investigated the buckling behaviour of symmetric balanced square laminated composite plates theoretically and experimentally. The obtained results showed that failure starts from the plate edges and develops along the plate width. Using a numerical study on $[\pm\theta]s$ and $[\theta/\theta+90]s$ layups, they showed that by increasing the θ from 0 to 90, the critical buckling load decreases continuously. Dawn [8] conducted a study on the buckling of laminated plates where he has considered different ply orientation sets with different thickness for each ply. Using a refined shear theory, he found that the buckling load is sensitive to the ply orientation and that the -/+45 ply orientation is the most sensitive to transverse shearing taking place in laminates plates. Langvarsky and Bocko [9] investigated the buckling of composite plates having different orientations of layers. The authors used finite element based analysis in Abaqus CAE. They considered four different orientations with three, six and twelve layers loaded by shear and compression loads. They identified the inappropriate orientations of layers for different type of loads. They noticed that the orientations of composite layers affect the value of the critical buckling load. In particular, the composite plates with three layers and orientation of layers [90,0,90] was found to be more sensitive to compression load than shear load.

Considering buckling of laminated composite plates with delamination, interaction between the delamination and the buckling strength exists. Growth of delamination was related to the total strain-energy release G by a technique based on a virtual crack closure, Whitcomb and Shivakumar [10]. The method enabled to predict a large variation of G along the delamination front for square and rectangular delaminations. Self-similar growth is then not expected to occur as a delamination may grow in the load direction or perpendicular to the load direction. This depends on the current delamination aspect ratio, the strain level, and the absolute size of the delamination.

In the framework of the following analysis, delamination is assumed to be stationary and not affected by the actual state of loading. The buckling behaviour of a stiffened composite panel is studied by means of finite element modelling under Abaqus software package. The panel is discretized by using special shell elements, which allows defining the material of composite panel, the thickness and layout of layers in the composite stiffened panel. We consider a carbon Epoxy Prepreg stiffened panel with different plies orientations sets. The initial distributed geometric imperfection is taken into account in order to assess its influence on the buckling strength. This has the form of the first Euler buckling mode which is determined by a pre-analysis. The geometrically nonlinear problem is solved by using Riks Algorithm. The limit load is identified as that for which the related load increment changes the sign.

2 Materials and methods

2.1 Geometry and initial imperfection

Geometric nonlinearities are considered while the material is assumed to have a linear elastic behaviour. The simulation is done by means of the Abaqus software and the famous S4R shell element is used. This element has four nodes with six degrees of freedom at each node. It can consider both material and geometric nonlinearities. In the present analysis, the geometrical configuration of the stiffened panel is fixed and the boundary conditions that were considered in the numerical simulations are represented as shown in figure 1. The displacements are designated by u and the rotations by θ.

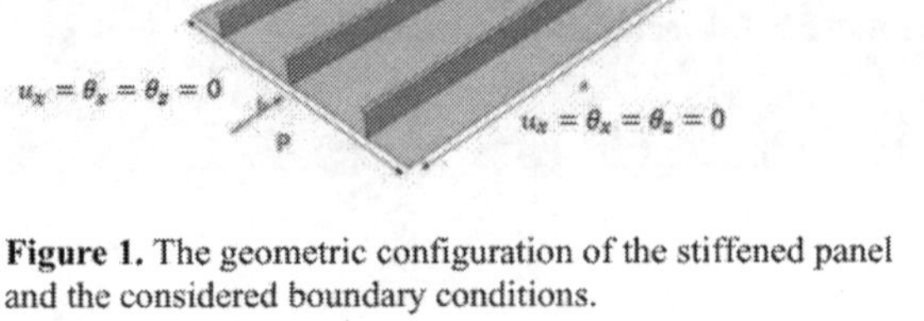

Figure 1. The geometric configuration of the stiffened panel and the considered boundary conditions.

The total length of the plate is $a = 958\,\text{mm}$ and the width is $b = 757.5\,\text{mm}$. The stiffeners take the shape of L with a fixed thickness $t_w = 2.95\,\text{mm}$ supposed uniform, their height is $h_w = 64\,\text{mm}$, the thickness of the flange is $t_f = 4.3\,\text{mm}$ and its height is $b_f = 12\,\text{mm}$, see figure 2.
The skin plate of the stiffened panel is supposed to be distorted by initial imperfections due to the assembling process. This distortion is assumed to be proportional to the first Euler buckling mode of the stiffened panel. This is provided by performing a linear buckling analysis. The following eigenvalue problem is solved to get the buckling load multipliers λ_i and buckling modes ψ_i

$$([K]+\lambda_i[S])\{\psi_i\}=0$$

where [K] is the stiffness matrix and [S] is the stress stiffness matrix. The stress stiffness matrix is computed based on the initial stress state induced by the applied load.
The first buckling mode is obtained by using an algorithm available under Abaqus that extracts the buckling modes ψ_i associated to the previous eigenvalue problem.
The initial geometric imperfection is then defined by taking the maximum amplitude of distortion equal to w_0. The first bucking mode is given in figure 2.

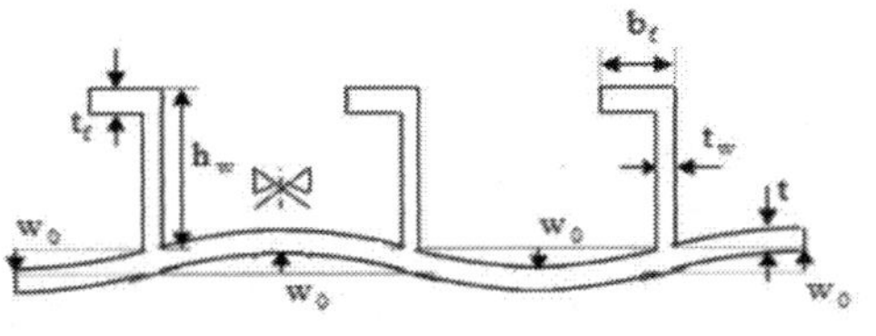

Figure 2. Geometric parameters of the panel and characteristics of the initial distortion in the xy plane

2.2 Material properties

The material properties considered are those of a carbon epoxy prepreg with the following properties: density $\rho = 1450\,\text{kg}/\text{m}^3$, Poisson's coefficient $\upsilon=0.15$ and rigidity modules of the orthotropic elastic matrix $E_1=139\,\text{GPa}$, $E_2=35\,\text{GPa}$, $G_{12}=G_{13}=50\,\text{GPa}$ and $G_{23}=42\,\text{GPa}$.
The composite is made of various plies that have different orientations θ. The thickness of each layer is given as 0.0015 m. In the skin plate four symmetric layers were used and three symmetric layers for the stiffeners.
The orientations of plies in the composite are given in table 1.
Figure 3 shows an example of the orientation set of the fibres in the composite as obtained from Abaqus CAE. The set of plies is denoted [0,90,45,-45]s, where s stands for symmetric

Table 1. The orientations sets of the plies in the composite used in the simulations

Set of plies Orientations	Skin	Stiffeners
1	[45,0, -45,90]s	[45,0,45]s
2	[0,90,45, -45]s	[45,0,45]s
3	[45,0, -45,90]s	[45,0,45]s
4	[30, -30, -30,30]s	[45,0,45]s
5	[45,0, -45,90]s	[30, -30,30]s
6	[0,90,45, -45]s	[30, -30,30]s
7	[45,0, -45,90]s	[-30, -30,30]s
8	[30, -30, -30,30]s	[30, -30,30]s

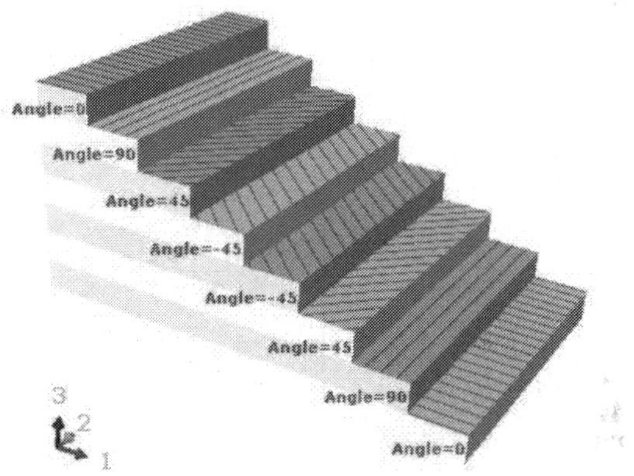

Figure 3. The orientation of the fibres in the skin of the composite in the case of [0,90,45, -45]s.

2.3 Buckling strength analysis

For a given combination of parameters including sets of plies orientations and amplitude of initial imperfection, a finite element model is built under Abaqus software packages. The modelling integrates geometric nonlinearities and iterations are controlled using the Ricks method. This type of analysis is used usually to study the instability and the post-buckling behaviour of the structure and solves both the load and the displacement during the analysis using another variable which is the arc length. The buckling load is determined to be that one associated with the limit point on the curve which gives the resulting axial load as a function of the axial shortening of the panel.

3 Results and discussion

To assess the influence of initial geometric imperfections on the buckling strength of composites panels, three levels for the modal imperfection amplitude were considered: $w_0 = 2, 4, 6\,\text{mm}$. Different plies orientation sets were also considered as given in Table 1.
The buckling strength load for given plies orientations and initial geometric imperfections are summarized in Table 2.

Table 2. Buckling strength load of the composite panel as function of the plies orientations and initial imperfection amplitude

Plies Set	w_0=2 mm	w_0= 4mm	w_0=6mm
1	7115025	6954494	6571126
2	7168455	6830670	6448324
3	7473821	6909395	6520073
4	7012424	6695725	6315012
5	7473821	7099268	6607864
6	7402314	6997066	6522682
7	7476491	7109686	6640345
8	7261103	6816493	6322702

The results show that the orientations of the layers and the initial geometric imperfection have important effect on the buckling strength of the composite panel. Increasing the amplitude of the initial imperfection yields decrease of the buckling strength load. The most adverse combination of the orientations of layers for the composite stiffened panel which is associated to set number 4: [30,-30,-30,30]s for the skin plate and [45,0,45]s for the stiffeners. This was the case for the three values of initial imperfection amplitudes. The most favourable combination of the layer's orientation is given by the set number 7: [45, 0, -45, 90]s for the skin plate and [-30, -30, 30]s for the stiffeners. Moreover, the buckling shapes were found to be slightly affected by the orientations and the numbers of layers of the composite panel.

Conclusion

To avoid the early collapse of the composite stiffened panel, it is preferable to take various cautions when designing the structure, in choosing the type of material and the geometry. In this work, the buckling strength of a composite stiffened panel loaded in compression was studied by using the finite element method. The effects of the different layers orientations of the fibres in the composite and the initial geometric imperfection resulting of the assembling process were taken into account. It was found that the buckling strength depends slightly on the orientations of the fibres and largely on the amplitude of initial geometric imperfections. Optimisation of the plies orientations is then possible.

References

1. A.W. Leissa, ASME Appl. Mech. Rev. **40**, 575 (1987)
2. Y.X. Zhang, C.H. Yang, Composite Structures 88, 147 (2009)
3. F. Cappello, D. Tumino, Composites Science and Technology **66**, 264 (2006)
4. A. Karrech, M. Elchalakani, M. Attar, A.C. Seibi, Composite Structures **174**, 134 (2017)
5. K. Liang, Q. Sun, Composite Structures **168**, 266 (2017)
6. O. Namdar, H. Darendeliler, Composites Part B **120**, 143 (2017).
7. M. Heidari-rarani, S.S. Khalkhali-sharifi, M.M. Shokrieh, Comput. Mater. Sci. **89**, 89 (2014)
8. C. Dawn, *Effects of Thickness and ply orientation on buckling of laminated plates* (NASA Tech. Memo. 87691, 1986)
9. P. Lengvarský, J. Bocko, M. Hagara, Am. J. Mech. Eng. **4**, 413 (2016)
10. J.D. Whitcomb, K.N. Shivakumar, Journal of Composite Materials **23**, 714 (1989)

Detection of rebar corrosion in bridge deck by using GPR

Tahar Bachiri[1], Abdellatif Khamlichi[2,a] and Mohammed Bezzazi[1]

[1]*Department of Physics, Faculty of Sciences and Technology at Tangier, University Abdelmalek Essaadi, Tangier 93001, Morocco*
[2]*Department STIC, National School of Applied Sciences at Tetouan, University Abdelmalek Essaadi, Tetouan 93030, Morocco*

Abstract. Ground Penetrating Radar based inspection technique is currently used for mapping bridge deck in order to monitor the health of this vital structure. Research over recent years in this field has focused on identification of rebar corrosion which is known to cause staged deficiencies by reducing bridge deck strength to critical levels. It is expected that using high-frequency antennas could improve discrimination of anomalies and enhance detection of small corrosion defects affecting bridge deck rebars. In this work, simulation of the input reflection coefficient of reinforced concrete bridge deck as modified by corrosion of a steel rebar is performed. The objective is to analyse the effect of the work frequency and corrosion layer thickness on the input reflection coefficient. Simulations have been carried out for various situations involving intact and corroded zones of a rebar buried in a reinforced concrete plate with the whole infrastructure overlaid by an asphalt layer. The obtained results have shown that the value of work frequency is very important and should be carefully selected.

1 Introduction

Ground Penetrating Radar (GPR) utilizes the characteristics of an electromagnetic field in the time domain and samples the backscattered echoes from anomalies that can be present in the host medium. GPR is nowadays used in different applications such as Non-Destructive Testing based techniques for inspection of civil engineering infrastructures [1].

Principle current applications of GPR in the field of civil engineering include determination of construction features, estimation of element thickness, estimation of the concrete cover depth, location of reinforcing bars and metallic ducts, location of corrosion variations, localization of voids and estimation of rebar size [2,3].

GPR has been used since the 1970's to inspect various forms of civil infrastructures, and seems to have nowadays the widest appeal in bridge deck inspection. This method is in competition with infrared thermography which has been in use since the late 1970's and which became commercially viable two decades ago. Following recent improvements in testing equipments, infrared thermography has turned out to be interesting as a bridge deck inspection tool. However, this technique still suffers from its inability to infer the depth of a delaminated or deteriorated area.

GPR systems record the response to a radar impulse or a series of radar impulses incident on the surface of the test part. The acquired reflection echoes are processed to obtain a radargram either in the form of A-scan, B-scan or C-scan. Careful assessment of the GPR responses can often yield information about the material properties and its condition. Although anomalies can be evident in the GPR scanned image of an object, their quantification and interpretation in terms of position and main features is an issue that remains today quite confusing. As a result, all civil infrastructure defects cannot be diagnosed reliably with a single sensing modality, and inspections can just identify problems once they have progressed to a certain grade of severity.

Studying the waveforms and patterns generated during GPR scanning in terms of radargram forms constitutes the key approach towards interpretation of results. These can be presented in a variety of ways including A-scan which shows the reflected signal waveform as function of two flight travel time. Reflected signal peaks by anomalies are then quite visible on the obtained scaled curve. Associating these traces to the characteristics of anomalies is still a huge challenge that necessitates considering in general solution of an inverse problem.

The primary cause of deterioration in bridge decks is due to concrete cracking which frequently results in delamination like defects. Corrosion of steel reinforcements over time constitutes also a major concern in bridges. This can be accelerated by de-icing salts and water infiltration. Furthermore, when the reinforcements corrode they expand and a horizontal crack is formed in the reinforcement level [4]. Most of applications have focused on detecting corrosion induced deteriorated areas in bridge decks [5].

Condition assessment and system management of Bridge decks have been carried out by using GPR for quantifying concrete overlay above reinforcing steel, pavement thickness, and concrete permittivity to assess concrete quality. GPR units were also found to be

[a] Corresponding author: khamlichi7@yahoo.es

capable of identifying concrete with high chloride and moisture contents since these factors produce high variable reflections of the radar wave at the overlay-deck interface [7]. Changes of the wave reflected by concrete-rebar interfaces have been used to identify rebar characteristics and to assess corrosion probability [8]. Bridge decks overlaid with asphalt do not pose a problem for GPR mapping of underlying reinforced concrete infrastructure since asphalt consists principally of aggregates that behave as a dielectric medium for electromagnetic wave propagation.

Both ground-coupled and air-coupled antennas can be used in bridge deck surveys. Ground-coupled systems can provide very detailed information about the bridge deck's structures and reinforcement bars. The major problems are the slow speed of data collection and that bridge lanes have to be closed during the data collection. This is why the use of air-coupled antenna systems, which can perform data collection without causing major traffic problems, is highly recommended, especially on high traffic volume roads.

In a bridge deck survey, high frequencies are needed to achieve high resolution as required by identification of small anomalies. As propagation velocity of electromagnetic waves in concrete varies typically from $0.1 m/ns$ to $0.12 m/ns$, using a GPR with the standard frequency $f=1.5GHz$ enables theoretically a vertical resolution of at least $40mm$, while $f=2GHz$ yields a resolution exceeding the range of $30mm$ [9]. Detecting rebar corrosion in a bridge deck was performed successfully by GPR having a ground coupled antenna working at the frequency of $1.5GHz$ [8]. Another study conducted on a bridge deck in Finland demonstrated that the use of a GPR antenna with a frequency of $1GHz$ resulted in good radar images [4].

In this work, inspection is considered for reinforced concrete bridge deck overlaid with asphalt by the input reflection coefficient. This consists of a global coefficient resulting from complex sequences of interactions, involving multiple scatterings from the rebar-concrete and the asphalt-concrete interfaces and also from corroded zones in rebars or other defects taken inside the concrete. Focus is on bridge deck suffering from a corroded rebar with various extents. A simplified modelling of corrosion gain is considered here for the reflection from the corroded rebar. A parametric study is conducted after that in order to determine under which circumstances it is possible to operate detection of a corrosion band that is present at the surface of a rebar buried in concrete.

2 Materials and methods

GPR can detect echoes from all three types of electrical parameters contrasts: the permittivity ε in F/m, the permeability μ in H/m and the conductivity σ in Ω/m. The permeability is normalized with respect to the free space permeability: $\mu_0=4\pi\times10^{-7} H/m$, such as $\mu=\mu_r\mu_0$, while the permittivity is normalized with respect to the permittivity of free space $\varepsilon_0=8.8541878176\times10^{-12} F/m$ and the relative permittivity as: $\varepsilon_r=\varepsilon/\varepsilon_0$. From wave propagation point of view, it is motivating to introduce complex relative dielectric constant under the form

$$\varepsilon_r=\varepsilon_r'-j\varepsilon_r'' \tag{1}$$

where ε_r' is the real part of the complex permittivity, ε_r'' is the imaginary part of the complex permittivity and $j=\sqrt{-1}$. In general, the real part is responsible for the velocity with which waves travel, while the imaginary part controls the attenuation experienced by the wave.

Under harmonic excitation at frequency f, the relative dielectric constant writes

$$\varepsilon_r=\varepsilon_r'-j\frac{\sigma}{2\pi f\varepsilon_0} \tag{2}$$

The amplitude and polarity of the wave reflected from a change in dielectric properties depend on the ratio of the two relative dielectric constants and is referred to as the reflection coefficient. Assuming a plane wave normally incident from medium 1 to 2, the reflection coefficient $R_{1,2}$ can be expressed as follow

$$R_{1,2}=\frac{\sqrt{\mu_{r1}\varepsilon_{r2}}-\sqrt{\mu_{r2}\varepsilon_{r1}}}{\sqrt{\mu_{r1}\varepsilon_{r2}}+\sqrt{\mu_{r2}\varepsilon_{r1}}} \tag{3}$$

When the magnetic permeabilities of the two materials are equal to that of free space, Equation (3) can be expressed in term of only the dielectric constants of the top (1) and bottom (2) media as

$$R_{1,2}=\frac{\sqrt{\varepsilon_{r2}}-\sqrt{\varepsilon_{r1}}}{\sqrt{\varepsilon_{r2}}+\sqrt{\varepsilon_{r1}}} \tag{4}$$

The magnitude of the reflection coefficient dictates the amplitude of the reflected wave, and the sign of the reflection coefficient dictates the phase shift of the returning wave. Knowing the common values of concrete dielectric properties, reflection occurring on carbon steel is almost quasi-total: $R_{1,2}\approx 90\%$.

From Equation (3), it is possible to obtain the power reflection coefficient as

$$Pr=(R_{1,2})^2=\left(\frac{\sqrt{\mu_{r1}\varepsilon_{r2}}-\sqrt{\mu_{r2}\varepsilon_{r1}}}{\sqrt{\mu_{r1}\varepsilon_{r2}}+\sqrt{\mu_{r2}\varepsilon_{r1}}}\right)^2 \tag{5}$$

Air-coupled GPR produces more difficult data to work with as transmission is poor, but the antennas are not in contact with the ground and data can be collected at a much higher speed without damaging the antennas. For air-coupled GPR, the problem of bridge deck made from reinforced concrete and overlaid with asphalt can be assumed to involve a stratified medium consisting of N homogeneous layers. The upper layer boarding the deck upper side is air and the layer at the bottom side is associated to steel rebar level which can be supposed to be a perfectly reflecting surface. An analytic formulation of the reflection and transmission coefficients can be developed by considering the bridge deck configuration geometry depicted in Figure 1. A uniform plane wave with either TE or TM polarisation is emitted by a horn like antenna located at the distance d_0 from the deck surface called the leading interface which serves as the reference for the reflection and transmission coefficients.

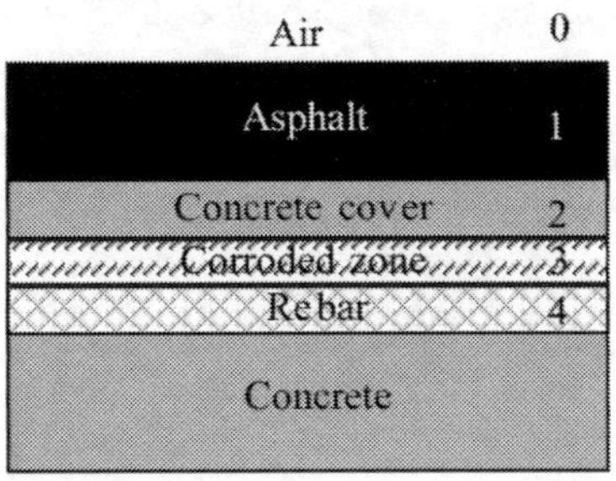

Figure 1. Bridge deck layers with perfect reflecting rebar layer.

The overall reflection and transmission coefficients for perpendicular polarization can be determined recursively [10]. Considering the layers described in Figure 1 under normal incidence, then the value of N=3 and the corroded zone is labelled 3. For this particular situation, one obtains the input reflection coefficient according to the following four recursive steps:

$$Y_4=\sqrt{\frac{\varepsilon_4\mu_3(1-j\tan\delta_4)}{\varepsilon_3\mu_4(1-j\tan\delta_3)}}$$

$$A_3=\frac{\exp(d_3\gamma_3)(1-Y_4)}{2},\quad B_3=\frac{\exp(-d_3\gamma_3)(1+Y_4)}{2}$$

$$Y_3=\sqrt{\frac{\varepsilon_3\mu_2(1-j\tan\delta_3)}{\varepsilon_2\mu_3(1-j\tan\delta_2)}}$$

$$A_2=\frac{\exp(d_2\gamma_2)}{2}\left[A_3(1+Y_3)+B_3(1-Y_3)\right]$$

$$B_2=\frac{\exp(-d_2\gamma_2)}{2}\left[A_3(1-Y_3)+B_3(1+Y_3)\right]$$

$$Y_2=\sqrt{\frac{\varepsilon_2\mu_1(1-j\tan\delta_2)}{\varepsilon_1\mu_2(1-j\tan\delta_1)}} \qquad (6)$$

$$A_1=\frac{\exp(d_1\gamma_1)}{2}\left[A_2(1+Y_2)+B_2(1-Y_2)\right]$$

$$B_1=\frac{\exp(-d_1\gamma_1)}{2}\left[A_2(1-Y_2)+B_2(1+Y_2)\right]$$

$$Y_1=\sqrt{\frac{\varepsilon_1\mu_2(1-j\tan\delta_1)}{\varepsilon_0\mu_1}}$$

$$\Gamma_{in}=\exp(-2d_0\gamma_0)\frac{A_1(1-Y_1)+B_1(1+Y_1)}{A_1(1+Y_1)+B_1(1-Y_1)}$$

with for n=1,...,4

$$\gamma_n=2\pi f\sqrt{-\mu_n\varepsilon_n+j\frac{\sigma_n\mu_n}{2\pi f}} \quad \text{and} \quad \tan\delta_n=\frac{\sigma_n}{2\pi f\varepsilon_n'}$$

where d_n is the thickness of layer number n, Γ_{in} is the input reflection coefficient and T_{in} is the input transmission coefficient.

The reflection coefficient Γ_{in} contains information on the layers crossed by the electromagnetic wave. It depends in particular on the thickness of the corroded zone. The presence of corrosion modifies the input reflection coefficient Γ_{in} through the coefficients A_3, B_3, Y_3 and Y_4.

In the absence of corrosion, the layer number three vanishes and the recursive steps are modified according to

$$Y_4=\sqrt{\frac{\varepsilon_4\mu_2(1-j\tan\delta_4)}{\varepsilon_2\mu_4(1-j\tan\delta_2)}}$$

$$A_2=\frac{\exp(d_2\gamma_2)(1-Y_4)}{2},\quad B_2=\frac{\exp(-d_2\gamma_2)(1+Y_4)}{2} \qquad (7)$$

$$Y_2=\sqrt{\frac{\varepsilon_2\mu_1(1-j\tan\delta_2)}{\varepsilon_1\mu_2(1-j\tan\delta_1)}}$$

with the last step (last four lines) remaining the same as in equation (6).

3 Results and discussion

In order to assess the effect of corrosion, a Matlab script was developed for both corrosion being present in the bridge deck, equation (6), and corrosion free situation, equation (7). Comparison was then made between these two conditions as function of wave frequency.

A bridge deck having the form given in Figure 1 was considered. The problem parameters are given in Table 1, where some conductivies are function of the wave frequency. The investigated interval of frequencies was fixed to be $[1,2]GHz$ and corrosion zone thickness was taken in the interval $[0,5]mm$. Figure 2 gives the evolution of the input reflexion coefficient as function of the work frequency for the corrosion free bridge deck an three different values of the corrosion layer thickness.
Figure 3 gives the input reflexion coefficient as function of the corroded layer thickness for various values of the work frequency. Note that there are intersection between the curve x=0 which corresponds to a corrosion free bridge deck and any other amount of corrosion thickness. This occurs for a specific frequency.
Figure 2 shows that, for a given thickness of the corrosion layer, there are frequencies for which no distinction can be made in terms of the input reflection coefficient between corroded rebar layer and corrosion

 https://doi.org/10.1051/matecconf/201819100009

free bridge deck. Figure 3 shows that for f=1.75GHz and for two different corrosion levels the input reflexion coefficient has the same value. This signifies that estimation of the amount of corrosion cannot be made by working with this frequency.

Table 1. Geometric and dielectric properties of the bridge deck layers as function of the work frequency f and corrosion layer thickness x .

Layer	$d\,(cm)$	μ_r	ε_r'	$\sigma\,(S/m)$
Air	1	1	1	0
Asphalt	2	1	4.18	0
Concrete	3	1	6.20	$2\times10^{-11}\times f$
Corrosion	x	$2-0.3j$	1.5	$5.56\times10^{-12}\times f$
Rebar	$2.5-x$	100	1.45	9.931×10^{6}

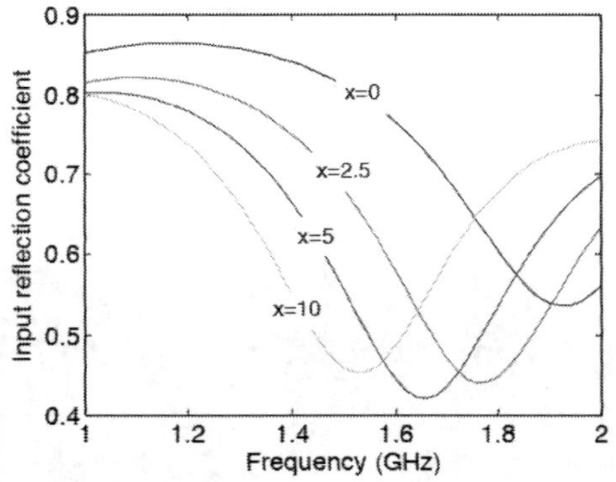

Figure 2. Input reflection coefficient as function of the harmonic wave frequency for various corrosion layer thicknesses.

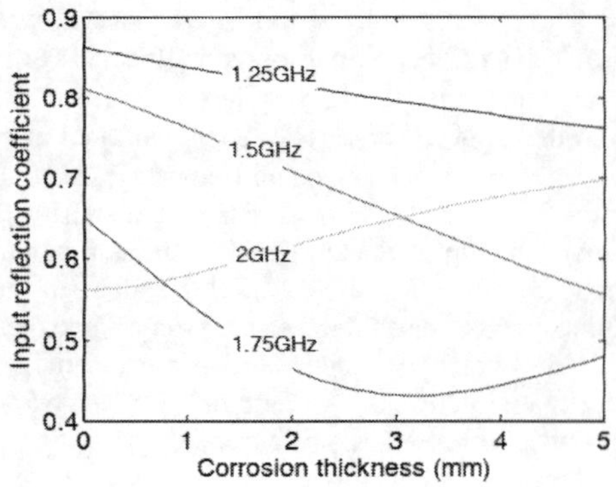

Figure 3. Input reflection coefficient as function of the corrosion layer thickness for various harmonic wave frequencies.

The obtained results emphasize the fact that for operating corrosion detection in bridge deck, one has to choose the right frequency, otherwise corrosion can be masked or corrosion extent cannot be estimated. As the bridge deck configuration is variable, it is recommended in practice to use a packet of frequencies.

4 Conclusions

A methodological approach for the detection of rebar corrosion anomaly in bridge deck made from reinforced concrete was presented. This methodology employs GPR acquired data in terms of the overall input reflection coefficient when perpendicular polarization of the harmonic electromagnetic wave is used under normal incidence. It was found that the input reflection coefficient depends hugely on the frequency of harmonic wave. As selection of the wave that enables to monitor corrosion depends on the actual problem, it is recommended then to use a packet of waves centred on a given high frequency. This will increase resolution and corrosion detection. Experimental validation is needed in order to assess these findings.

References

1. H. M. Jol, *Ground Penetrating Radar: Theory and Applications* (Elsevier, Amsterdam, 2008)
2. J.H. Bungey, Constr. Build. Mater. **18** (2004)
3. V. Barrile, R. Pucinotti, NDT&E International **38** (2005)
4. N. Parry, J.L. Davis, Geological Survey of Finland **16** (1992)
5. T. Alongi, C.G. Clemena, P.D. Cady, *Condition Evaluation of Concrete Bridges relative to Reinforcement Corrosion* (SHRP Report, Washington, DC, 1992)
6. I. Al-Qadi, K. Jiang, S. Lahouar, *Bridge deck reinforcing steel cover depth prediction using GPR* (Transportation Research Board 85th Annual Meeting, Washington, 2006)
7. K.R. Maser, K. Tabrizi, V. Gangi, *Use of Nondestructive Methods for Large Scale Bridge Deck Evaluation* (TRB, National Research Council, Washington, 2001)
8. S. Laurens, J.P. Balayssac, J. Rhazi, G. Klysz, G. Arliguie, Materials and Structures **38** (2005)
9. C.W. Chang, C.H. Lin, H.S. Lien, Constr Build Mater, **23** (2009)
10. C.A. Balanis, *Advanced Engineering Electromagnetics* (Wiley, New York, 1989)

Sensitivity of the inverse problem solution related to detection of rebars buried in concrete by using GPR scanning

Zoubaida Mechbal[1,a] and Abdellatif Khamlichi[2]

[1] *Department of Physics, Faculty of Sciences at Tetouan, University Abdelmalek Essaadi, Tetouan 93030, Morocco*
[2] *Department STIC, National School of Applied Sciences at Tetouan, University Abdelmalek Essaadi, Tetouan 93030, Morocco*

Abstract. Ground Penetrating Radar is a device which is nowadays largely used in civil engineering applications. Considering a rebar buried in a concrete medium, this work addresses sensitivity of the inverse problem solution associated to identification of the object radius and its depth from B-scan data acquired by the GPR. The approach uses a closed form parameterisation of the hyperbola trace emerging in the radargram as function of the hyperbola apex coordinate along the direction of B-scan, the cover depth, the radius of the object and the relative permittivity of the medium. Estimation of the wave velocity, the hyperbola apex coordinates and the rebar radius was performed through solution of an appropriate nonlinear least mean squares problem. Perturbation analysis was then conducted by assuming that the hyperbola points coordinates, extracted from raw data of radargram, are randomly distributed according to Gaussian densities of probabilities. The effect of the amount of data was also analyzed. The method was implemented in Matlab environment. The obtained results have shown that identification process is extremely sensitive to noise affecting the B-scan raw data, but not to the number of points used in identification.

1 Introduction

Ground Penetrating Radar (GPR) is commonly used in civil engineering for the localization of supply lines, land mines, pipes and many other buried objects in soils. Nowadays, its use encompasses diagnosis and detection of rebars in reinforced concrete structures. One of the changeling issues in this filed is the localization and the identification of radius of buried steel bars.

Detection of a reinforcement steel rebar can be made from the raw data recorded as reflection hyperbola trace in the GPR radargram. The hyperbola has typical shape that varies according to the depth and material of the rebar as well as the surrounding medium dielectric properties. One of the most important problems to deal with in the field of GPR based detection techniques is to retrieve from the actual radargram the hyperbola characteristics by post-processing of acquired B-scan data and inverse problem solution.

The hyperbola characteristics enable to identify the centre position and the radius of the buried cylindrical object. Several methods have been developed in recent years in order to automate this task [1-3]. The various authors who dealt with this matter have used essentially three different approaches in order to perform image processing techniques of the obtained radargrams, Ristic et al. [4]. These include Hough transform based techniques [5-6], interactive visual overlapping of hyperbola [7] and direct fitting of the raw data to a general hyperbola equation [8-9].

It was recognized that the two first approaches suffer from critical drawbacks as the error affecting the radius estimation of buried cylinder radius may be in some cases as large as 50%, because of numerical bad conditioning that impedes correct localisation of hyperbola centre [8]. However, the third approach was judged to be more interesting as it allowed reducing the computational cost with higher accuracy.

In the following the third approach according to [4] is considered. This method is based on mathematical representation of the raw data of radargram by means of a parameterized hyperbola which is strained then to fit the observed trace hyperbola as formed by reflected amplitudes from rebar obstacle. The first work in this context was performed by Shihab et al. [8]. The authors have given an estimation of the radius by assuming prior knowledge of the velocity from measurement. They proposed a hyperbola fitting technique using weighted least squares with respect to the mathematical error expectation of two-way travel time.

Dolgiy et al. [10] have presented several techniques for optimal radius estimation of subsurface metal pipes. These included the classical weighted least squares, the recursive Kalman filter, the maximum likelihood, the direct least-square fitting of hyperbola and the Nelder-Mead optimization. Ristic et al. [4] have proposed a multi-step nonlinear least squares fitting procedure in order to simultaneously estimate velocity and radius of buried cylinders. The method was implemented in a Matlab environment and validated by means of

[a] Corresponding author: mechbal.zoubaida@gmail.com

experimental results. The new method was found to be significantly more accurate and robust with regard to both noise and the amount of raw data.

In this work, a closed form parameterisation of the hyperbola trace that is acquired in the B-scan radargram is considered as function of the hyperbola apex coordinate along the direction of B-scan, the cover depth, the radius of the object and the wave velocity in the medium. Use is made of an appropriate least squares based method in order to assess more effectively sensitivity of the inverse problem solution associated to identification of rebars buried in homogeneous concrete. This approach to identification problem is recognized to be fast enabling to perform large parametric studies to deal with the effects of the amount of data used and uncertainties affecting the trace hyperbola points.

To take into account uncertainties, probabilistic analysis is conducted. The coordinates of hyperbola points, extracted from raw data of radargram, are assumed to be randomly distributed according to Gaussian densities of probabilities. Then, sensitivity of the obtained estimated parameters to the number of points used in identification and inherent uncertainties in the raw data is analyzed.

2 Materials and method

2.1 Inverse problem solution

The hyperbolic signatures in GPR data are created by reflections that occur on the target boundary due to changing distance between the antenna and the target.

Let's assume a B-scan for the target having the form of a disk of radius R and centre with the coordinates, $(x_0, t_0+2R/v)$ where v is the wave velocity in the host medium, x_0 is the location of the antenna position at the nearest distance from the scanning surface and t the two-way travel time, see Figure 1. If the antenna is located at the abscissa x, then the following geometrical relation holds [12]

$$\left(\frac{v}{2}t+R\right)^2=(x-x_0)^2+\left(\frac{v}{2}t_0+R\right)^2 \quad (1)$$

The hyperbola equation can be rewritten under the form

$$t=t(x_0,t_0,v,R,x)=-\frac{2R}{v}+\frac{2}{v}\sqrt{\left(\frac{v}{2}t_0+R\right)^2+(x-x_0)^2} \quad (2)$$

Denoting N the total number of points used and for a given set of data points $(t_i,x_i)_{i=1,\dots,N}$ located on the hyperbola which mathematical model is defined by equation (2), the unknowns (x_0,t_0,v,R) can be determined by a least squares based method as solution of the following minimisation problem

$$(x_0,t_0,v,R)=\underset{(\xi,\tau,\zeta,\rho)}{Argmin}\sum_{i=1}^{N}\left[t(\xi,\tau,\zeta,\rho,x_i)-t_i\right]^{2\alpha} \quad (3)$$

where α is a parameter that fixes the cost function to be minimized.

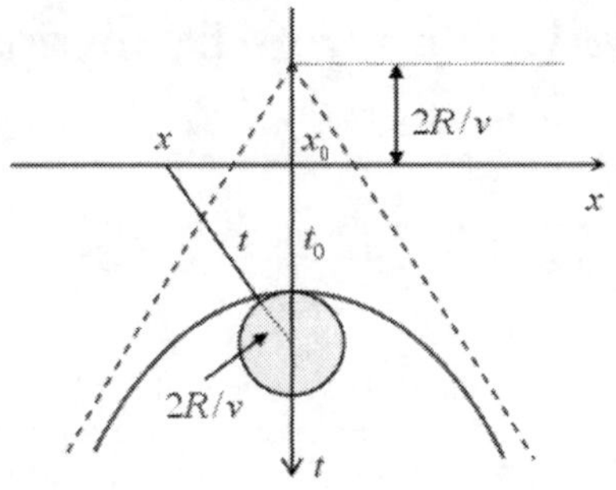

Figure 1. The GPR hyperbola trace signature showing the problem parameters (x_0,t_0,v,R).

The previous problem can be readily solved by using suitable minimization algorithms. The algorithm based on pattern search method for linearly constrained problems proposed by Lewis and Torczon [2] proved to be very efficient. This algorithm can be applied through the command *patternsearch* available in Matlab. In order to stabilize the inversion problem, it was suggested [1] to add the following bound constraints associated to the domain where the solution is expected to be found:

$$\begin{bmatrix}0\\0\\c/10\\0.004\end{bmatrix}\le\begin{bmatrix}x_0\\t_0\\v\\R\end{bmatrix}\le\begin{bmatrix}L\\20L/c\\c\\0.04\end{bmatrix} \quad (4)$$

where L is the maximum dimension of the concrete member being monitored and c the speed of wave in air. In equation (4), the lengths are expressed in m, time is in s and the velocities are in m/s.

The inverse problem solution depends like this, among other things, on the total number of points N which is fixed a priori at the start of the procedure.

2.2 Probabilistic analysis

If the raw data acquired by GPR during a B-scan is perfect, then the solution of the optimization problem defined by equations (3) and (4) will be only affected by the errors associated to the numerical procedure used. But, in practice the obtained radargram suffers from various sources of noise. The acquired data points $(t_i,x_i)_{i=1,\dots,N}$ are perturbed and their actual positions may be far for the theoretical ones. To take into account this uncertainty, it is assumed here that these points are random variables with their coordinates distributed according to Gaussian densities of probabilities with uniform standard deviations in space and time. In practice, the real distributions of probabilities can be identified from experimental testing followed by standard statistics performed on the results obtained from a large number of samples.

Probabilistic analysis is performed in this work by using simulations with raw data perturbed by generated random numbers. The uncertainty affecting hyperbola points is assumed to be distributed according to Gaussian density of probabilities. The Matlab command *normrnd* which relies on the Matlab random number generator *rand* is used.

Fixing the standard deviations in time σ_t and in space σ_x, simulations enable to quantify evolution of the errors on the identified hyperbola parameters as modified by the perturbed raw data.

3 Results and discussion

Let's consider a reinforced concrete specimen containing a buried rebar of radius $R=4mm$, see Figure 2. The concrete section has the form of a rectangle having the edge dimensions $1m$ (horizontally) and $0.8m$ (vertically). These dimensions are taken to be large enough in order to avoid any reflection from the bottom edge of the block during the selected time interval.

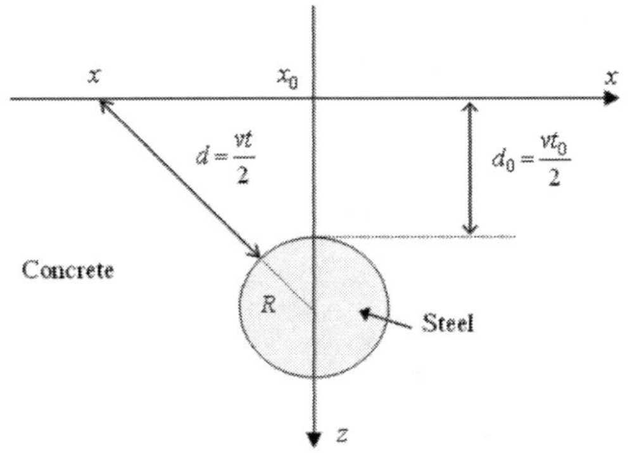

Figure 2. Geometrical characteristics of a buried steel rebar in concrete.

The dielectric properties chosen for the different materials are given in Table 1.

Table 1. Materials properties used for Reflexw simulation

Material	ε_r	μ_r	σ
Concrete	4	1	0.0255
Steel	1.45	1	9.931×10^6

The work frequency was fixed at $1.5GHz$, the cover depth d_0 was taken to be $0.52m$ and x_0 was fixed at $0.5m$. The wave speed is given by $c/\sqrt{\varepsilon_r}=1.5\times10^8\,m/s$ and $t_0=6.9333\,ns$.

For the direct problem simulations associated to the previous data have been performed by using Reflexw software package [11]. A B-scan which consisted in moving at constant speed the antenna along the x-line was performed. The simulation parameters under Reflexw were selected as follows: the wave form excitation was according to Kuepper pulse; the source type was exploding reflection; the boundary conditions were of linear absorbing range type; the output type was single line which corresponds to B-scan, the range of time was fixed at $15ns$. The obtained radargram is depicted in Figure 3.

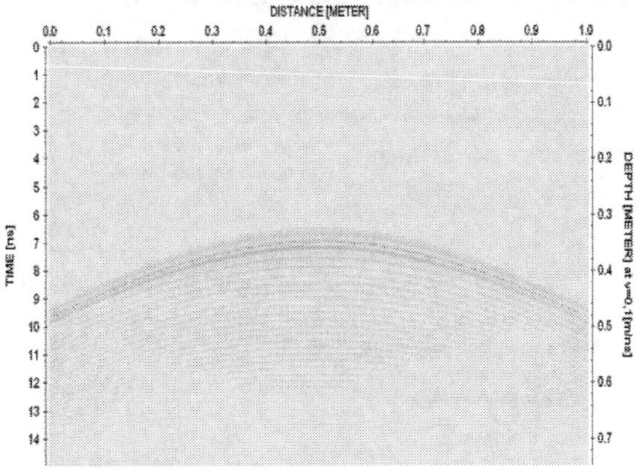

Figure 3. The obtained image radargrams corresponding to rebar have diameter $R=4mm$

Figure 3 shows the hyperbola trace on the obtained B-scan radargram. The raw data points located on the trace hyperbola are used to solve the optimisation problem as defined by equations (3) and (4) through using the command *patternsearch* in Matlab. Denoting X the vector which components are the unknowns (x_0,t_0,v,R), the following instructions were used in Matlab:

```
LB=[0,0,c/10,4.e-3];
UB=[L,20*L/c,c,4.e-2];
xx0=(LB+UB)/2;
options = psoptimset('TolMesh',1.e-12,'MaxIter',1000,
'TolX',1.e-12,'TolFun',1.e-12,'MaxFunEvals',5000);
for k=1:10
[X,fval,exiflag]=patternsearch(@(X)Zt_object(X,xi,ti),
xx0,[ ],[ ],[ ],[ ],LB,UB,options);
xx0=X;
end
```

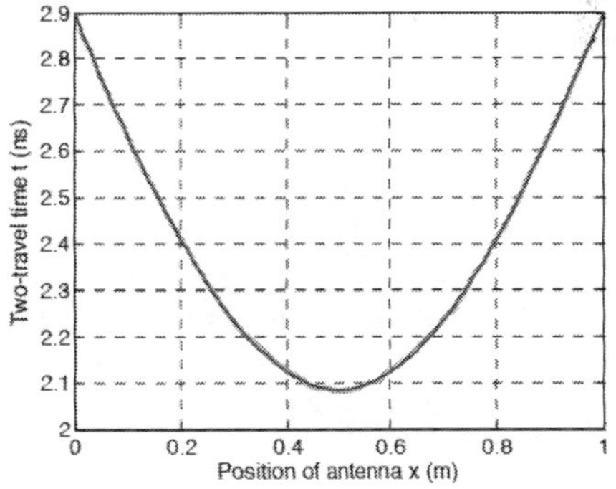

Figure 4. The identified hyperbola in blue and the input hyperbola corresponding to raw data in pink; $\alpha=2$ and $N=64$.

Figure 4 shows for $\alpha=2$ the identified hyperbola, by the procedure indicated above, superposed to the hyperbola resulting from the radargram raw data. Even if graphically the two curves are almost identical, the hyperbola centre which is also fixed by the hyperbola asymptotes is here inaccurate. Bad convergence in terms of the hyperbola characteristics is in fact obtained when $\alpha=1$ or $\alpha=2$. The relative error for the rebar radius is too large $\Delta R/R \approx 800\%$, while the other errors remain small: $\Delta x_0/x_0 \approx 0.06\%$, $\Delta t_0/t_0 \approx 0.3\%$ and $\Delta v/v \approx 4\%$.

On the opposite, excellent convergence was obtained for integer values of the cost function parameter α that belong to the interval $[3,10]$. The obtained relative errors are in this case: $\Delta x_0/x_0 \approx 0.01\%$, $\Delta t_0/t_0 \approx 0.1\%$, $\Delta v/v \approx 1\%$ and $\Delta R/R \approx 0\%$.

Fixing $\alpha=10$, the number of points fixing the raw data hyperbola was reduced from $N=64$ to respectively $N=33$, $N=17$, $N=9$ and $N=5$.

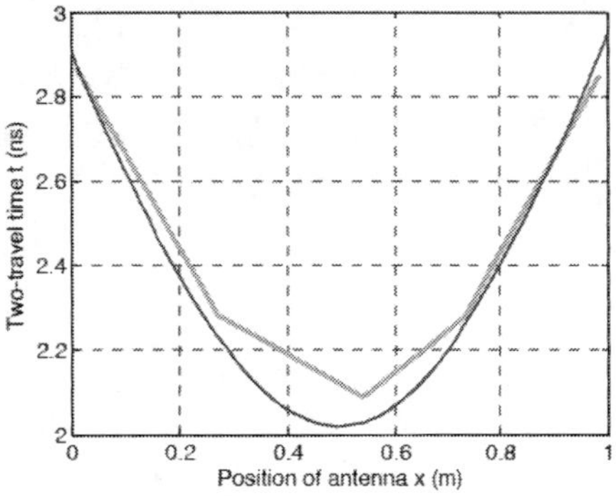

Figure 5. The identified hyperbola in blue and the input hyperbola in pink; $\alpha=10$ and $N=5$.

Figure 5 shows the obtained results for $\alpha=10$ and $N=5$. The obtained relative errors are: $\Delta x_0/x_0 \approx 1.3\%$, $\Delta t_0/t_0 \approx 2.9\%$, $\Delta v/v \approx 6\%$ and $\Delta R/R \approx 0\%$.

Fixing $\alpha=10$, perturbation of the raw data of the hyperbola trace given in Figure 3 was performed by sampling according to Gaussian density of probabilities. Use was made of the Matlab function *normrnd* which generates random numbers from the normal distribution with given mean parameter and standard deviation. Here the means are assumed to coincide with the initial input raw data, while the standard deviations are taken to be: $\sigma_t = 0.0015$ and $\sigma_t = 0.001$. This choice represents only small perturbations of the input raw data which is limited to a concentrated zone around the initial points. Performing a total number of 100 simulations, the error affecting rebar radius identification is shown in Figure 6.

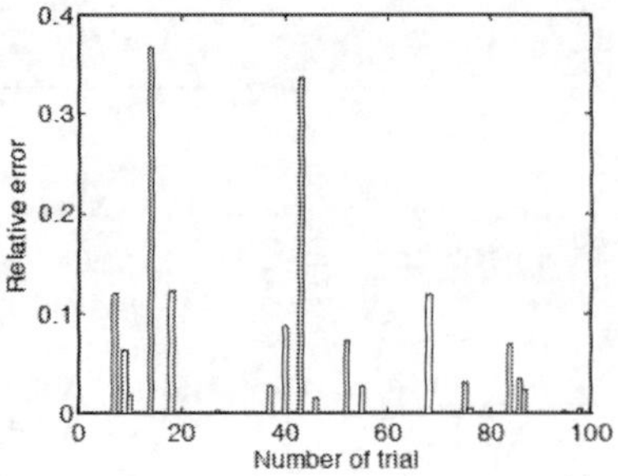

Figure 6. The error on the identified rebar radius for $\alpha=10$ and $N=64$, $\sigma_t = 0.0015$ and $\sigma_t = 0.001$.

The results regarding the influence of parameter α on convergence have shown that choosing large values of α enables to reduce the influence of accidental points and makes it possible to take into account more effectively contribution of normal points to the fitting hyperbola. Accuracy is then improved.

The results associated to the effect of the number of points taken on the raw data hyperbola have shown that these can be reduced favourably to a small number of points that are regularly spaced.

Figure 6 has shown that even small perturbation of the raw data points has considerable consequence on the identified rebar radius as the error was large.

4 Conclusions

In this work, sensitivity of the inverse problem solution associated to identification of rebar in concrete was analysed. The inverse problem was stated as an optimisation procedure enabling to search for the trace hyperbola characteristics that fits the raw data provided by a B-scan radargram. Solution of the inverse problem was fixed by appropriate selection of the cost function. Then random perturbation was applied to the raw data in order to guess robustness of the procedure. It was found that even small deviation of raw data points as modelled by a Gaussian density of probabilities yields significant false identified rebar radius. Reduction of the number of points was found to not affect largely the identification error if these are regularly spaced.

References

1. Z. Mechbal, A. Khamlichi, NDT&E International, **89** (2017)
2. R.M Lewis, V. Torczon, SIAM Journal on Optimization, 10 (2000)
3. A. Vaso Ristic, D. Petrovacki, M. Govedarica, Computers & Geosciences **35** (2009)
4. A.V. Ristic, D. Petrovacki, M. Govedarica, Computers & Geosciences **35** (2009)
5. C.G. Windsor, L. Capineri, INSIGHT, Journal of the British Institute of Non Destructive Testing **5** (1998)
6. G. Borgioli, L. Capineri, P.L. Falorni, S. Matucci, C.G. Windsor, IEEE Transactions on Geoscience and Remote Sensing **46**, 8 (2008)
7. G.R. Olhoeft, Journal of Applied Geophysics **43** (2000)
8. S. Shihab, W. Al-Nuaimy, A. Eriksen, *Proceedings 10th International Conference on Ground Penetrating Radar* (Delft, The Netherlands, 2004)
9. S. Shihab, W. Al-Nuaimy, Subsurface Sensing Technologies and Applications **6** (2005)
10. A. Dolgiy, A. Dolgiy, V. Zolotarev, *Proceedings of 11th International Conference on Ground Penetrating Radar* (Columbus Ohio, USA, 2006)
11. K.J. Sandmeier, *Reflexw version 6.0* (Karlsruhe, Germany, 1998)

https://doi.org/10.1051/matecconf/201819100011

Infrared thermography used for composite materials

Abdelaziz Mouahid[1,a]

[1]*SCD Laboratory, National School of Applied Sciences at Tetouan, University Abdelmalek Essaadi, Tetouan, Morocco*

Abstract. Many areas of the industry use composite materials, because of their good mechanical features in terms of low density and high mechanical strength. Composite materials are used wherever elevated rigidity and strength with reduced unit weight are required; such as wind turbine blades, shipbuilding, aeronautical and aerospace. However, the properties of composites can be hugely affected because of inside defaults such as delaminations or local cracks. Several non-destructive methods have been used for the verification of defects during construction or operation, such as ultrasound or x-ray. These methods are costly and difficult to implement. Non-destructive method using infrared thermography is considered very useful and works perfect with low cost. Two methods of non-destructive detection by infrared exists, which are (i) passive thermography, that consists of measuring infrared stream emitted by the material and (ii) active thermography, which consists of heating the material and measuring the cooling of material surface using an infrared camera. This communication describes the basic principles of both passive and active thermography, and then describes other different methods for detection of composite materials defects.

1 Introduction

Infrared thermography is a technique that allows non-destructive control of a composite material by an infrared camera.

Composite materials are made by assembling of at least different materials. The first material called the reinforcement is mostly in form of fibers, which enhances mechanical strength. The second material called matrix ensures the cohesion of the structure and transmission of the efforts towards the reinforcement.

Composite materials provide good mechanical performance with a low density and are widely used in the industry like wind turbines, aeronautical, naval and space. Composites are largely used in blades and nacelles of wind turbines. Blades have to be manufactured with strong and lower weight material in order to decrease the cost of the installation. The use of carbon / glass fiber allows producing higher stiffness and lower weight blades [1]. Figure1 shows the cross section design of a wind turbine blade. The blade design incorporates pre-impregnated epoxy glass fiber and carbon fiber [1].

In shipbuilding industry, steel has been used for many years and was the dominating building material. Recently, composite materials are being used in large scale. Composite materials offer good mechanical features with weight reduction which results in decreased fuel consumption. Safety is increased in case of fire and total cost is lowered. Fiber glass Reinforced Plastic (FRP) (which is a composite material made of a polymer matrix reinforced with fibers) is like this increasingly used in ship building industry [2].

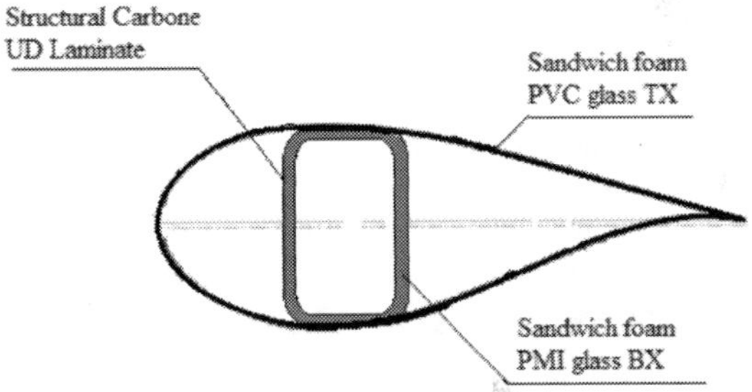

Figure 1. Cross section design of wind blade.

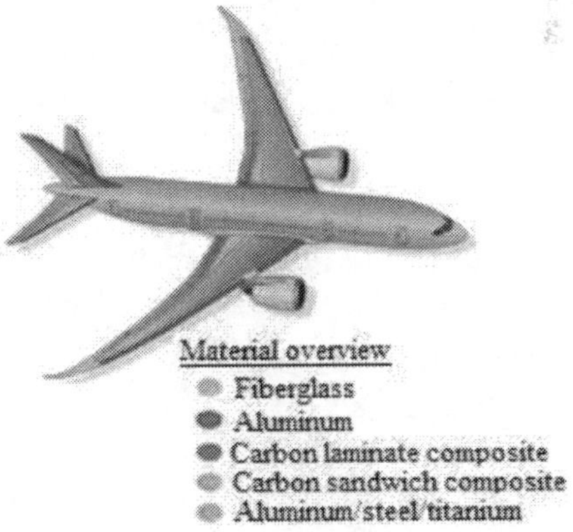

Figure 2. Composite materials used in 787 Boeing airplane [4].

In aerospace industry, the use of composite represents 50% of the materials of which the majority is

[a] Corresponding author: mouahidaziz@gmail.com

made up of carbon fiber composite [3]. Figure 2 shows composite materials used in 787 aeroplane.

Composite material properties are largely affected by the damage of their structure. Thermography can detect different types of composite damage such as delamination, cracks, etc. Figure 3 give more details of some composite defaults.

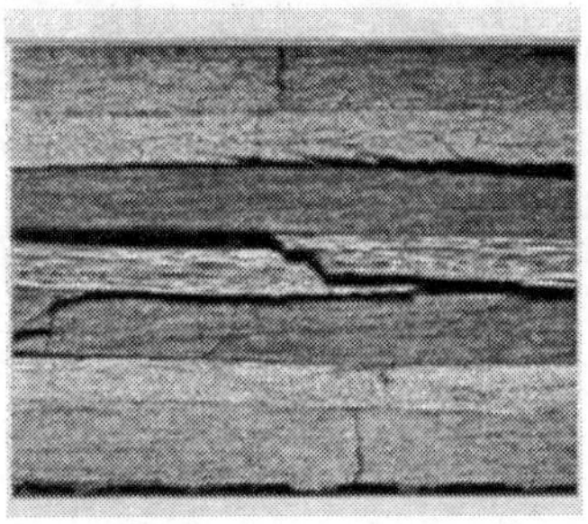

Figure 3. Breaking and delamination in composites [5].

2 Infrared waves

Any solid material with a temperature above absolute zero (-273.15°C) emits in permanence energy in the form of electromagnetic radiation. The movement of electrically charged particles traveling through vacuum or matter produces this energy. Figure 4 shows categories of electromagnetic spectrum based on wavelength.

Infrared radiation is an electromagnetic wave of frequency lower than visible light, and a wavelength between the visible area ($\approx$ 0.7 μm) and the field of the microwave ($\approx$ 1 μm).

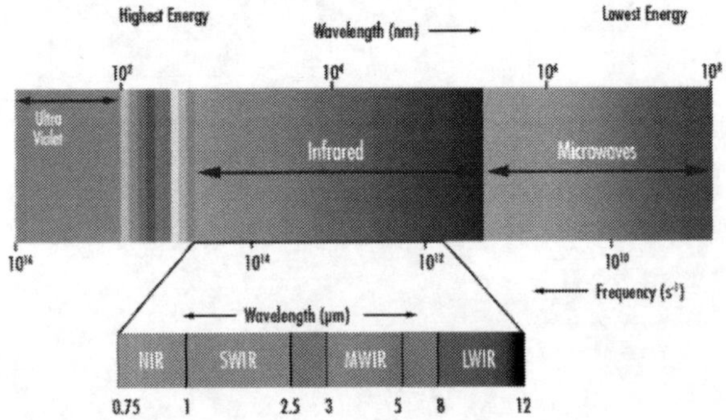

Figure 4. Infrared electromagnetic spectrum.

Infrared radiation can be subdivided further into three categories:

- **Near infrared:** with wavelength between 0.75 μm and 1.5 μm detected by specialized photographic emulsions (up to 1 μm), photoconductors detectors and photovoltaic.
- **Mid infrared:** with wavelength between1.5 μm and 20 μm detected by thermal, photoconductors, and photovoltaic detectors
- **Far infrared:** with wavelength between 20 μm and 1000 μm detected only by thermal detectors.

Special camera captures infrared signals, with optics made of germanium, allowing focusing the heat radiation.

Detection can be made with either cooled infrared cameras that is very sensitive or bolometric camera with good quality/price.

Thermal imaging taken by infrared camera may contain errors because:

- Some surfaces are more reflective than others;
- Emissivity of some objects can influence camera's temperature calculation;
- Values should be adjusted to room temperature;
- Viewing angle should be adjusted.

Infrared thermography is widely used in many areas such as:

-Wind turbine industry where infrared is used for detecting defects on blades made of composite materials during manufacturing and operation;
- Night vision (army, or navigation...) or search for persons during fires;
- Civil engineering for insulation evaluation, search for points of loss of heating;
- Electrical and mechanical construction or maintenance;
- Medicine: detection of hyperthermia in humans (in airports...);
- Veterinary medicine: detection of localized inflammation (tendonitis...);

For non-destructive detection of composite materials defects, two techniques are used:

- Passive infrared thermography ;
- Active infrared thermography.

3 Passive infrared thermography

This technique is used to measure naturally occurring temperature contrasts in the composite or induced natural solicitations (Sun, temperature of the air...). A thermal camera is used and captured images are treated by computer (see Figure 5).

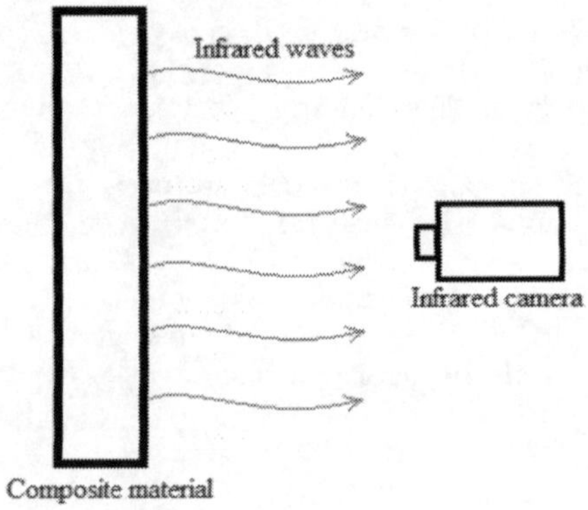

Figure 5. Passive infrared thermography.

Defects covered by this technique are superficial or in few millimeters deep. Images can be captured at any

https://doi.org/10.1051/matecconf/201819100011

moment for superficial defects or to be received at different times.

The advantage of this passive method is that it is simple to use, and measurements can be made on large surfaces. This method is limited and presents weaknesses because measures are influenced by the room temperature, images can contain the shadow of the objects on the surface, and treated depth of the material is only few millimeters.

4 Active infrared thermography

Composite sample is subjected to artificial stimulation and we measure induced temperature contrasts (see figure 6).

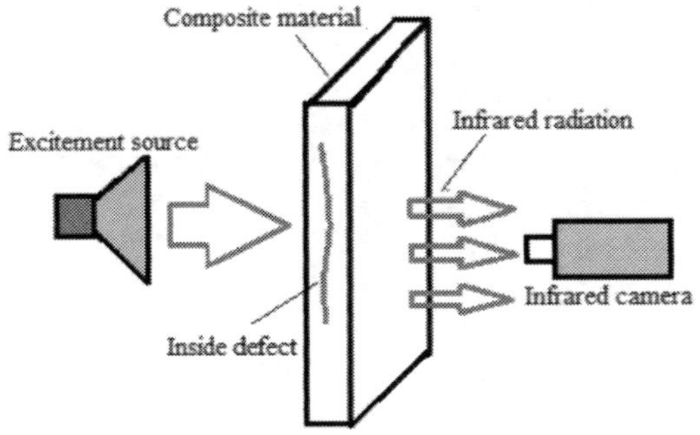

Figure 6. Active infrared thermography.

The source of excitement can be a flash, a halogen lamp, induction by electric current or vibro-thermography.

We use several methods for active thermography technique including:
- Pulsed thermography (PT);
- Lock in thermography (LT);
- Vibro-thermography stimulated by ultrasound (VTU);
- Laser infrared thermography (LT);
- Infrared induction thermography (IT).

4.1 Pulsed thermography (PT)

This technique consists of emitting a thermal pulse to the composite material (for instance: using lamp, laser, or any other heat source) and record the evolution of the temperature at the surface level in specific time interval (see Figure 7).

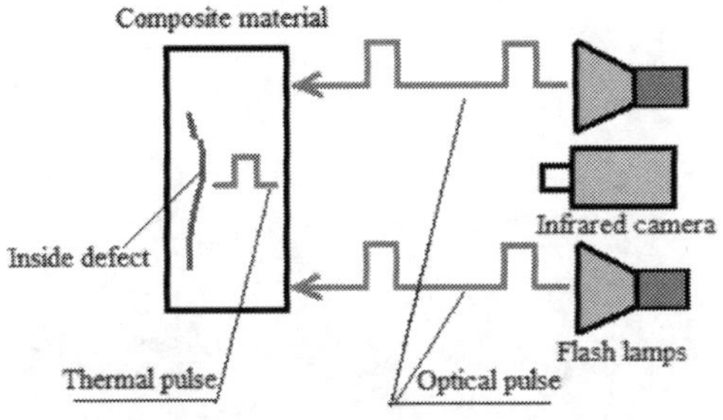

Figure 7. Schematic of pulsed thermography.

Thermal balance of the composite material is disturbed by providing a pulse with very concentrated and very brief energy. A powerful Xenon lamp can generate large thermal power on the composite material surface. Quick infrared camera can record successive images which can be analyzed in real-time scale pixels. The energy pulse sent to the composite material decreases due to thermal conduction. The reduction in temperature profile after pulse injection gives information about the properties of the material. This pulsed infrared thermography enables detecting flaws or determination of coatings characteristics.

The pulsed method has advantages such as it is not affected by the inhomogeneity of the composite heating and the surface properties that may be changing. However, the depth of investigation is limited.

4.2 Lock in thermography (LT)

Figure 8 illustrates LT method. This consists of submitting the composite material to a sinusoidal thermal solicitation with well-defined frequency, in steady state, and then capturing the surface heating using an infrared camera. The analysis covers the phase shift and amplitude of temperature response. LT gives information about structure and thermal inhomogeneity of composite material.

The apparatus of LT includes generator that provides modulated frequency and intensity to halogen lamps to produce heat waves. A high-resolution camera measures composite material thermal response. Analysis of phase shift between thermal excitation signal and thermal surface response allows locating defects if present in examined parts.

Lock in thermography is nondestructive control method that is recognized for evaluation of composite reinforced plastics carbon fiber in aeronautical, aerospace and automotive industry. It allows locating precisely defects and to determine their depth.

In addition, this method is extremely robust, insensitive to external disturbances and works even in difficult conditions. This method is ideal for quality control during production and operation.

The advantage of this technique is that it is not sensitive to nonconformities of heating, tests can be done on large surfaces, stimulation frequency can determine the depth of the test, and this method gives a good signal to noise ratio.

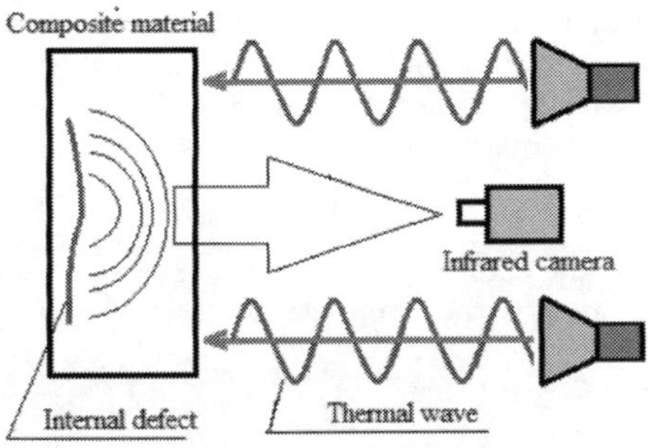

Figure 8. Schematic of Lock in thermography.

4.3 Vibro-thermography stimulated by ultrasound (VTU)

With ultrasound thermography or vibro-thermography, composite material is submitted to ultrasonic vibrations that propagate inside the material.

Test bench consists of generator with ultrasonic converter and infrared camera [5]. The heat emitted by the material surface is then captured using an infrared camera.

The interaction between thermal and mechanical waves can locate defects present in the composite material. They appear on the thermal images because they are releasing more heat by friction.

Vibro-thermography allows reaching deep flaws inside composite material, it also has the advantage of not being sensitive to the ambient temperature because ultrasonic waves propagate in depth and the surface of the material is not heated.

For the location of defects, the phase shift between excitement and thermal response is examined. Ultrasonic thermography gives information on the different types of defects such as composite cracks, detachments or delamination.

4.4 Laser infrared thermography (LT)

A laser beam is sent to the inspected composite material in order to produce heat on its surface [6]. A Well-defined area on the composite surface heated could be line or point.

Laser excitement creates heat that flow inside the composite material. Defects present in the composite material are considered as thermal barriers that affect the flow of heat. Infrared camera captures the thermal differences on the surface.

Composite material properties such as thickness, porosity and thermal conductivity can be detected using Fourier transform analysis.

Laser infrared thermography eliminates non-uniformity sensitivity of composite heating surface.

4.5 Infrared induction thermography (IT)

The principle of infrared induction thermography is to induce eddy currents within the composite material, and to capture the heat flowing from the surface using an infrared camera [7].

The loops of electrical current induced inside the composite material are uniform except when they encounter a defect in the material. The heat flow is then changed. During the test, variable magnetic field near the surface of composite material is created and the flow of temperature emitted by the surface of the material is measured.

The induction method is not sensitive to irregularities or impurities of the composite surface.

5 Conclusions

Composite materials are used in many areas such as wind energy, aeronautics, aerospace and marine construction. They offer good mechanical performance for low thicknesses.

Composite material manufacturing conditions greatly affect the quality of these materials.

Defects may appear during manufacturing or during operation, detecting these defects before at their earlier stage is then very important to ensure their quality and safety.

This article presented several non-destructive detection methods based on thermography techniques. A review of the most recent methods has been carried out. These methods have succeeded in locating and even in characterizing most of the defects that are present in composite materials.

Each technique has its advantages and drawbacks, the choice of a technique will depend on the type of defect to be detected, the degree of accuracy required and the cost of the equipment to use.

References

1. L. Mishnaevskyjr, K. Branner, H.N. Petersen, J. Beauson, M. McGugan, B.F. Sorensen, MDPI journal, **10** (2017)
2. D-A. Zisimopoulos, *Use of Fiber Reinforced Plastics in Ship Construction: A Study of SOLAS regulation II-2/17 on Alternative Design and Arrangements for Fire Safety*, (PhD Thesis, National technical university of Athens, 2015)
3. M. Mrazova, INCAS BULLETIN, **5** (2013)
4. S. Georgiadis, A.J. Gunnion, R.S. Thomson, B.K. Cartwright, Composite Structures, **86** (2008)
5. N. Montinaro, D. Cerniglia, G. Pitarresi, Composites Part B: Engineering, **108** (2017)
6. T. Li, D.P. Almond, D. Andrew, S. Rees, NDT & E International, **44** (2011)
7. B. Weekes, D.P. Almond, P. Cawley, T. Barden, NDT & E International, **49** (2012)

Robustness of ultrasonic detection of flaws by using synthetic aperture focusing technique

Sofia Rejani[1,a], Abdellatif Khamlichi[2] and Abdellah El-Hajjaji[1]

[1]*SCD Laboratory, Faculty of Sciences, University Abdelmalek Essaadi, Tetouan 93030, Morocco*
[2]*National School of Applied Sciences at Tetouan, University Abdelmalek Essaadi, Tetouan 93030, Morocco*

Abstract. One of the challenging problems in non-destructive evaluation is related to identification and sizing of flaws. A high resolution image of the scanned part is required. This allows, through using adequate post-processing of data, to perform localisation and sizing of a flaw. Several techniques have been introduced recently for this purpose. These include among others the synthetic aperture focusing technique, inverse wave-field extrapolation and the total focusing method. However, large uncertainties are affecting the inverse problem solution as provided by these methods when dealing with small defects. It was recognized that reconstruction based on the ultrasonic synthetic aperture focusing technique elaborated in frequency domain provides high resolution imaging even at large distances. This work focused on this promising procedure for the special case of ultrasonic imaging of flaws in 2D elastic medium under plane strain conditions, where the image is provided by a B-scan. Robustness of detection was investigated through perturbing the radargram by white noise and assessed as function of noise energy. It was found that synthetic aperture focusing technique is insensitive to noise.

1 Introduction

Flaws in parts can be produced during the manufacturing process or in service due to fatigue or induced accidental damage. In practice, it is important to determine the position of a flaw and estimate its size. Ultrasonic testing can be used for this purpose [1-2]. This may be performed by means of ultrasonic imaging which involves converting raw ultrasonic data collected by a transducer during a B-scan testing into an image representing with sufficient precision the interior state of the examined structure. This conversion can be achieved by the synthetic aperture focusing technique [3].

The concept of synthetic aperture appeared at first in the field of radar in 1951 when the technique of synthetic aperture radar (SAR) imaging was introduced [4]. This approach is based on a combination of measurements taken by moving radar to synthesize electromagnetic signals with a better resolution than that provided by fixed radar. The earlier algorithms dealing with synthetic aperture imaging were developed for SAR applications which are characterized by narrowband transmissions and very high propagation speeds [5]. Later on, the concept of synthetic aperture was proven to be relevant for side-looking sonar [6]. Then, the technique was applied in the field of medical imaging. In the 1970s, the synthetic aperture focusing reached the field of ultrasonics and his name became aperture focusing technique (SAFT) [7]. This flexible imaging method was firstly explored for non-destructive testing using ultrasonic transducers to inspect an isotropic medium. SAFT algorithm was used after that for reconstruction of defects in anisotropic media [8-9] where wave propagation is more complex because of the directional dependence of the material properties.

The SAFT method has been presented with both time domain and frequency domain versions. The frequency based version uses the Fast Fourier Transform (FFT) implementation, while the standard scanning ultrasonic testing has focused on the time domain. A special algorithm called delay and sum (DAS) was devised for this purpose [10]. This algorithm allowed obtaining an image with high resolution independently of the depth when compared with an ordinary B-scan imaging [11]. However, Olofsson has demonstrated the superiority of SAFT in frequency domain by using phased shift migration (PSM) technique [12]. Considering two experiments for immersed isotropic and layered objects in water, he concluded that the PSM is more efficient than the DAS algorithm. Moreover, the PSM algorithm has a lower computational complexity when compared with DAS algorithm. Advantage of the imaging by PSM has been also demonstrated in case of layered medium.

In this work, the SAFT in frequency domain, in its PSM version, will be applied for detection of flaws in a 2D isotropic elastic medium undergoing plane strains deformation. Simulation of B-scan testing is performed by using the Finite Element Method (FEM) under Abaqus software. Perturbation of the radargram is considered through introducing additive white noise in

[a] Corresponding author: Sofia.rejjani@gmail.com

the signal. Robustness of flaw detection is then investigated as function of the noise energy.

2 Materials and methods

The PSM method was first developed by Gazdag in 1978 for seismic records in inhomogeneous media [15]. Then, it was later adapted for ultrasonic imaging. This method is based on the exploding reflector model in which the point scatterers in the medium acts like a source emitting a pulse at $t=0$, and creating a wave field which is recorded by a transducer array. The exploding reflector model enables to consider the point scatterers in medium like point reflectors of the wave field. Besides, it neglects multiple reflections between layers and considers only upward-travelling waves.

2.1 Algorithm formulation

Let us consider an acoustic wave propagating in a multilayered medium which geometry is composed by L layers, numbered (1,...,L) in figure1. Each layer ℓ has two characteristics: the thickness of the layer, denoted d_ℓ, and the wave velocity, denoted c_ℓ. The depth z_ℓ represents the depth of the lower side of each layer ℓ.

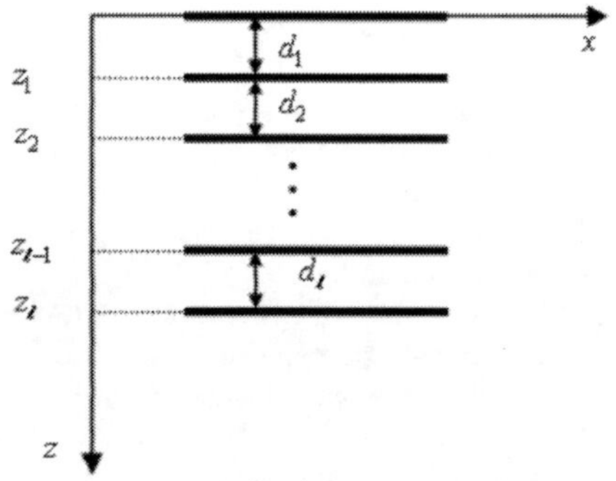

Figure 1. The multilayered geometry of the propagating medium

Let us denote by $p(x,z,t)$ the wave field emitted by the reflectors at $t=0$, and recorded by the transducer array in line $z=0$, this wave field can be extrapolated from the top line to others depths by decomposing it into upward-traveling waves of each layer of the medium and by superimposing it in order to find the final result. This emphasise the main idea encompassed in PSM method which intends extrapolating the observed wave field to create an image of the medium.

The expression of the upward-traveling wave at depth ℓ can be written as

$$p_\ell(x,z,t)=P_\ell(k_x,\omega)\exp\left[i\left(k_x x+k_{z,\ell}(z-z_{\ell-1})-\omega t\right)\right] \quad (1)$$

where $P_\ell(k_x,\omega)$ is the complex amplitude, ω the angular frequency, k_x and $k_{z,\ell}$ are the components of the wave number vector along the x-axis and z-axis.

Using the dispersion relationship

$$k_x^2+k_{z,\ell}^2=\frac{\omega^2}{c_\ell^2} \quad (2)$$

one obtains $k_{z,\ell}$ as a function of k_x and c_ℓ

$$k_{z,\ell}=-\sqrt{\frac{\omega^2}{c_\ell^2}-k_x^2} \quad (3)$$

with $\frac{\omega^2}{c_\ell^2}\geq k_x^2$ and the component $k_{z,\ell}<0$ because the wave travels upward along the z-axis.

Equation (1) can be rewritten as follows

$$p_\ell(x,z,t)=P_\ell(k_x,\omega,z)\exp\left[i\left(k_x x-\omega t\right)\right] \quad (4)$$

with

$$P_\ell(k_x,\omega,z)=P_\ell(k_x,\omega)\exp\left[ik_{z,\ell}(z-z_{\ell-1})\right] \quad (5)$$

Equation (4) represents the wave field emitted by the reflectors from the layer $\ell-1$ to the layer ℓ. Equation (5) gives the expression of the extrapolated wave field from layer $\ell-1$ to layer ℓ. We can generalize this equation to give the expression of the extrapolated wave field from the top layer 1 to any other layer ℓ in the medium, we obtain:

$$P(k_x,\omega)=P_\ell(k_x,\omega)\prod_{n=1}^{\ell-1}\exp(ik_{z,n+1}d_n) \quad (6)$$

One should notice that equation (4) represents the expression of a two-dimensional Fourier transform on x and t, which allows to pass from the temporal domain $p_\ell(x,z,t)$ to frequency domain $P_\ell(k_x,\omega,z)$. In equation (6) appears the phase shift which is represented by multiplication with the exponential involving the phase shift of all the layers above a given point at depth z. From equations (4) and (6), the phase shift migration PSM takes its name.

Considering the wave signal at $z=0$, one obtains then

$$p(x,z=0,t)=P(k_x,\omega)\exp\left[i\left(k_x x-\omega t\right)\right] \quad (7)$$

From which the amplitude $P(k_x,\omega)$ can be calculated by performing the two-dimensional Fourier transform on x and t. Then, by using equation (6), the amplitudes for a given layer can be calculated as

$$P_\ell(k_x,\omega)=P(k_x,\omega)\prod_{n=1}^{\ell-1}\exp(-ik_{z,n+1}d_n) \quad (8)$$

Using now equation (5), the amplitude at any depth z for a given wave frequency is obtained as

$$P_\ell(k_x,\omega,z)=P_\ell(k_x,\omega)\exp\left[ik_{z,\ell}(z-z_{\ell-1})\right] \quad (9)$$

By using equation (4) the wave signal at any depth z is

$$p_\ell(x,z,t)=P_\ell(k_x,\omega,z)\exp\left[i\left(k_x x-\omega t\right)\right] \quad (10)$$

https://doi.org/10.1051/matecconf/201819100012

Since in equation (10), the dependency on the layer is implicit, one can write simply

$$p(x,z,t)=P(k_x,\omega,z)\exp\left[i\left(k_x x-\omega t\right)\right] \qquad (11)$$

To obtain an image of the medium using PSM method, it is necessary to apply integration over ω in the frequency domain, followed by a 1-D inverse Fourier transform on x, at $t=0$ which indicates the time when the pulse is emitted from the reflectors. This corresponds to the time when a maximum concentration of the wave field occurs or in other terms the time that gives a focused image. The focused imaged pixel writes

$$p(x,z,t=0)=\int\int_{-\infty}^{+\infty} P(k_x,\omega,z)\exp(ik_x x)dk_x d\omega \qquad (12)$$

2.2 The imaging procedure by PSM

In summary, PSM imaging of a medium is done through the following procedure:

1. Calculate the two-dimensional Fourier transform of the radargram as obtained from a B-scan, where the transducer is moved along the x-axis for $z=0$ according to equation (7).
2. Calculate the wave field at the top layer interface using equation (8).
3. For each depth $z=z_\ell+\Delta z$ within the layer ℓ, shift the wave field by multiplying it with an exponential using (9).
4. Create a focused image by application of equation (12).

3 Results and discussion

3.1 Delamination detection by using PSM

Let us consider a plane strain elastic problem on a rectangular domain with the presence of a defect having the form of a cut-out as shown in Figure 2.

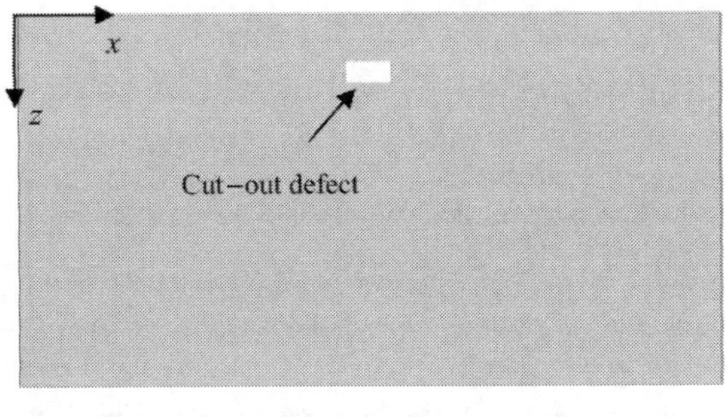

Figure 2. Plane strain elastic domain with a cut-out defect.

The rectangle has dimensions: length $a=300mm$, height $b=160mm$. The cut-out has the form of a rectangle which is centered in x- axis direction and located at depth $d=20mm$. The cut-out dimensions are length $a'=20mm$ and height $b'=10mm$. The defect size is chosen here large enough in order to avoid numerical discrepancies which could result from a very small defect. Here the objective is feasibility of the method with testing its robustness firstly for relatively large defects. This feature should also be tested for smaller flaws, but this issue is not considered in the actual study.

The material properties are density $\rho=2700kg/m^3$, Young modulus $E=70\,GPa$, Poisson coefficient $\nu=0.25$. These yield for the velocity of compressive waves $c_p=5577.7m/s$ and the shear waves velocity $c_s=3220.3m/s$.

The B-scan was achieved by selecting equally spaced points at uniform interval distance $\Delta x=4.688mm$ between points, beginning from the abscissa $79.687mm$ to the abscissa $220.312mm$, where the origin is selected on the left top corner of the rectangle. A total of 30 points were considered for excitation and recording at the same point the displacement along the z-axis direction. The excitation at a given point was a concentrated force having the form of a pulse, applied along the z-direction, and with the characteristics depicted in Figure3.

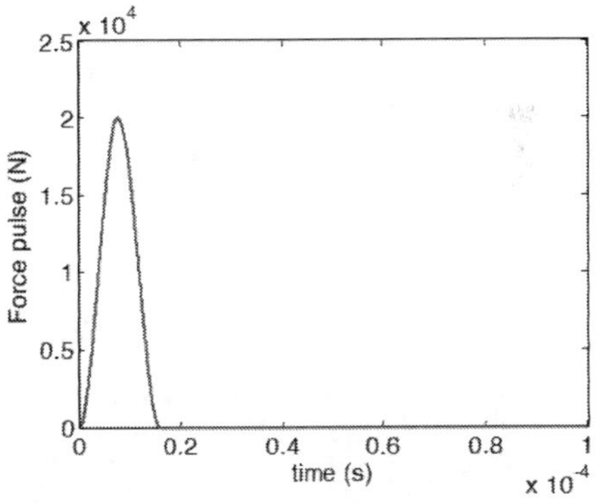

Figure 3. The pulse excitation vertical force

The step time was a user-defined increment of value $\Delta t_c=2ns$, the results are stored at the sampling time increment $\Delta t=1\mu s$, so the sampling frequency is $fs=1MHz$. The obtained raw data radargram is shown in Figure 4.

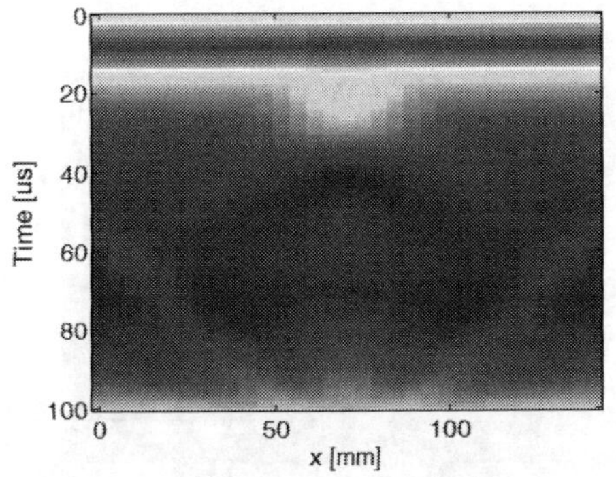

Figure 4. The raw data radargram before application of PSM.

The PSM algorithm is applied on the raw data in order to reconstruct the image of the rectangular domain with the cut-out defect. The obtained focused image is shown in Figure 5.

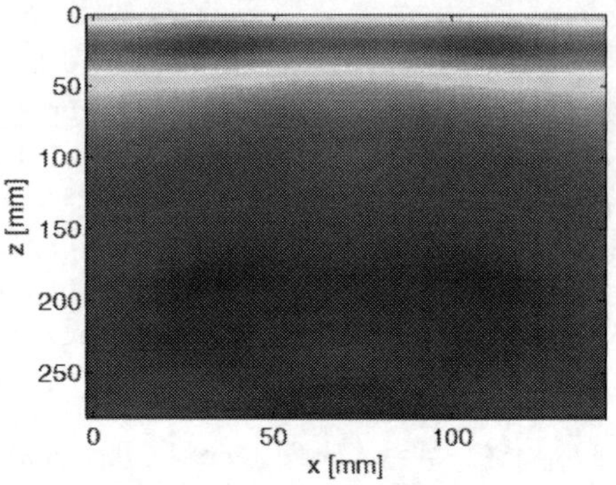

Figure 5. The focused image for literal data.

3.2 Robustness of detection by PSM algorithm

To assess robustness of PSM algorithm, the raw data was perturbed with a random noise. This noise affected each term of the raw data matrix, which in fact the tow-way travel time, with a pre-defined level of energy. Two levels of noise amplitude were studied: 20% and 100%. Figure 6 and Figure 7 presents respectively the focused image for 20% and 100% noise level.

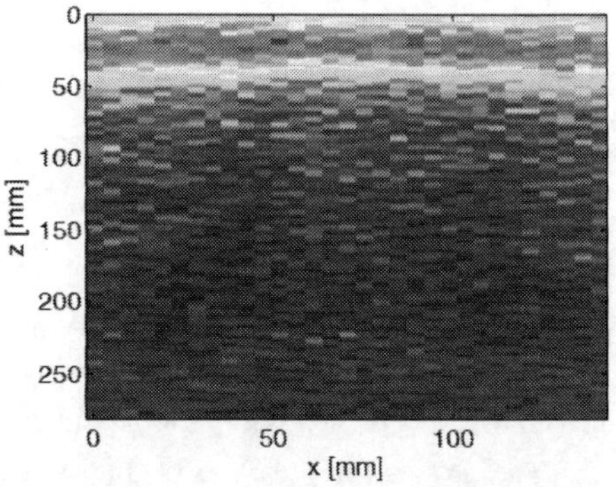

Figure 6. The focused image with 20% of noise influence.

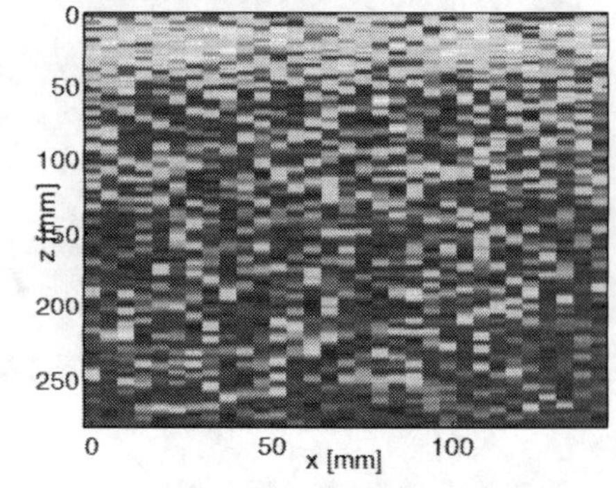

Figure 7. The focused image with 100% of noise influence.

From Figure 6, one can conclude that the algorithm is very robust in detecting defects as the defect signature is still clearly visible even for perturbations that have amplitudes reaching 20% of that of the acquired signal.

4 Conclusions

The synthetic aperture focusing technique was considered in this work in order to operate detection of flaws in two-dimensional elastic medium under plane stress deformation. Large uncertainties have been assumed to affect the raw data radargram acquired according to a B-scan. It was found that the synthetic aperture focusing technique in the frequency domain by using phased shift migration is robust against these perturbations. Further studies should be performed in order to assess this robustness feature as function of the flaw size and also its depth. Generalisation to composite structures should also be investigated.

References

1. A. Katunin, K. Dragan, M. Dziendzikowski, Composite Structures **127**, 1 (2015)
2. S. Li, T.P. Chu, IEEE Transactions on Ultrasonics, Ferroelectrics, and Frequency Control **60**, 2575 (2013)
3. C. Lane, *the development of a 2D ultrasonic array inspection for single crystal turbine blades* (Springer International Publishing, Switzerland, 2014)
4. C.A. Wiley, IEEE Transactions on Aerospace and Electronic Systems **AES-21**, 440 (1985)
5. L. Ferro-Famil, E.Pottier, *synthetic aperture radar imaging* (Microwave Remote Sensing of Land Surface, Elsevier, 2016)
6. M.P. Hayes, P.T. Gough, IEE Journal of Oceanic Engineering **34**, 207 (2009)
7. S.R. Doctor, T.E. Hall, L.D. Reid, NDT International **19**, 163 (1986)
8. R.A. Kline, *SAFT Imaging in Anisotropic Media* (Review of Progress in QNDE, Plenum Press, New York, 1999)
9. M. Spies, W. Jager, Ultrasonics **41**, 125 (2003)
10. M. Karaman, P.C. Li, IEEE Transactions on Ultrasonics, Ferroelectrics, and Frequency Control **42**, 429 (1995)
11. Q. Li, X. Yang, P. Huang, Y. Liao, Journal of Theoretical and Applied Information Technology **47**, 740 (2013)
12. T. Olofsson, IEEE Transactions on Ultrasonics, Ferroelectrics, and Frequency Control **57**, 2522 (2010)
13. J. Gazdag, Geophysics **43**, 1342 (1978)

Structural, optical and electrical characterizations of ZnO/PS

Zineb Yamlahi Alami[1,a], Moez Salem[2], Mounir Gaidi[2,3] and Jamal El khamkhami[1]

[1]*Condensed Matter Laboratory, Abdelmalek Essaadi University, Faculty of Sciences, M'hannech, Tetouan, Morocco*
[2]*Photovoltaic Laboratory, Research and Technology Centre of Energy, Borj-Cedria Science and Technology Park, Hammam-Lif, Tunisia*
[3]*Department of Applied Physics, University of Sharjah, Sharjah, United Arab Emirates*

Abstract. This work reported on solar cell ZnO/PS heterojunction fabricated from Zinc nitrate with molar concentrations 0.05M and 0.2M using spray pyrolysis technique, to study the effect of high and low molar concentration of Zn precursor on structural, optical and electrical properties. The structural analyse showed that the obtained thin films deposited on glass substrate were polycrystalline with a hexagonal wurtzite structure and preferentially oriented along the c-axis direction, while structural properties enhanced with higher morality 0.2M. Roughness surface of ZnO/PS HJ increased widely with higher morality. Through I-V characteristics, the enhancement of electrical properties with higher molarity has been achieved. I_{sc} get augment from 23mA/cm², to 27mA/cm², and V_{oc} from 551mV to 554mV when molarity has taken values 0.05M and 0.2M respectively. Ideality factor has influenced by molarity variance too. The impact of solar irradiance G and temperature T on solar cell fabricated from 0.2M molarity has been investigated. The current density increased from 13.53 mA/cm² to 27mA/cm², the voltage from 525mV to 546mV, and the efficiency from 5.2% to 10.9% when the solar irradiance have increased from 500 to 1000W/m². The temperature also influences on the solar cell behaviours, especially the voltage is enhanced by temperature increasing.

1 Introduction

Several works have focused on studying the n-ZnO/p-Si heterojunction behaviors because of its uses in different fields. For example, Chebil et al. [1] have deposited ZnO films on p-type crystalline silicon (p-Si) with [001] orientation, etched silicon and porous silicon by using sol gel spin coating method, they confirmed that the electrical properties of the ZnO layer deposition on porous are enhanced. Zebbar et al. [2] have prepared heterojunction structures of n-ZnO/p-Si by ultrasonic spray method, they obtained improved structural and optical properties as temperature of substrate increases. The ideality factor of obtained heterojunction was larger than *2*, the activation energy of saturation current was about *0.14eV* and the junction built-in potential deduced from C–V measurements was equal to *1.14V* at room temperature.

In this work, the effect of both molar concentrations *0.05M* and *0.2M* on structural and morphological properties of ZnO thin films nanoparticle deposited on glass substrate are studied. A layer of n-ZnO fabricated with both molarities, *0.05M* and *0.2M*, and allowed to grow on porous silicon substrate to fabricate a heterojunction solar cell is considered. The effect of deposing a layer of ZnO thin film nanoparticle with lower and higher molar concentration on I-V characteristics is studied. Then, factors ideality are extracted from I-V curves. Due to its high optical and electrical properties, the n-ZnO/p-Si fabricated by higher molar concentration *0.2M* is simulated using Matlab software. The objective is to study the impact of solar irradiance *G* between *500w/m²* and *1000w/m²* and cell temperature between *25°C* and *35°C* on behavior of this device.

2 Experimental work

Experiments were carried out on p-type, boron-doped multi-crystalline silicon substrate with a thickness of *350µm* and a resistivity of *0.5-2 Ω cm* [3]. ZnO films are prepared by the spray pyrolysis method using solution containing zinc nitrate [Zn $(NO_3)_2$, $2H_2O$] in *25ml* of deionized water. The concentration of the precursor solution was fixed at one of the values *0.05M* and *0.2M*.The spray rate of the [Zn $(NO_3)_2$, $2H_2O$] solution was kept constant at *2 ml/min* onto a preheated porous silicon at *500°C* temperature for *10min*.

The crystallographic structure of the films was studied by X-ray diffraction XRD technique. The surface morphology was characterized via atomic force microscopy (AFM). The reflectivity spectra of the different treated mc-Si surface was measured using a LAMBDA 950UV/Vis/NIR spectrophotometer equipped with an integrating sphere. The configuration for I–V measurements was performed using two-electrodes, which were applied on front and back extremities of the sample using Au and Al contacts, respectively. Current voltage measurements were performed using a computer

[a]Corresponding author a.yamlahizineb@gmail.com

controlled set up comprising a Keithley 220 current source and an Agelint 34401A multimeter under *1.5AM* illumination solar spectrum (*100 mW/cm²*).

3 Results and discussion

3.1 Structural studies

Figure1 shows the XRD of ZnO films prepared by *0.05M* and *0.2M*. Both these molarities exhibited an orientation along c-axis (002) perpendicular to the substrate surface, which indicates a hexagonal Wurtzite structure. However, the intensity of (002) plan has increased and 2θ position has shifted, taking *34.40°* then *34.51°* values respectively. For the ZnO powder, the peak position is 2θ=*34.42°*, the shift of the (002) peak position from its powder value are mainly associated with strain produced within the film [4].

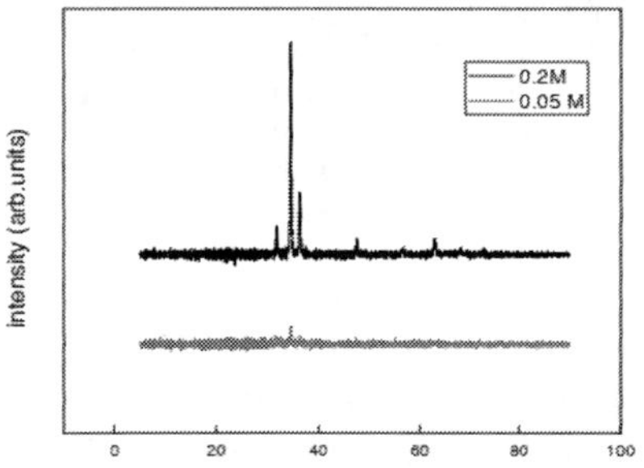

Figure 1. X-ray diffraction spectra of ZnO thin films deposited on glass substrate with different molar concentrations.

To have a clue about the crystallinity of the films, the crystallite size is calculated by Scherer formula:

$$D = \frac{0.9\lambda}{\beta \sin\theta} \quad (1)$$

where β is the full width half maximum of the peak (FWHM), λ is the X-ray wavelength (*1.5406Å* for CuK_{α}) and θ is the Bragg's angle.

The crystallite size increased from *10nm* to *33nm* for *0.05M* and *0.2M* respectively. We have calculated the dislocation density δ of the films by the formula [5]:

$$\delta = \frac{1}{D^2} \quad (2)$$

with D is the average grain size.

The dislocation density is defined as the length of dislocation lines per unit volume of the crystal. This parameter is taking $1\times10^{-2}nm^{-2}$ and $9.4\times10^{-4}nm^{-2}$ values for *0.05M* and *0.2M* respectively. Larger D and smaller δ values mean better crystallization of the films which indicates that the higher molar concentration reduces the crystal lattice imperfections.

To estimate the stress in the plane of the substrate surface, the c-axis strain ε_{zz} values has been calculated using the following equation [6]:

$$\varepsilon_{zz} = \frac{c - c_0}{c_0} \times 100 \quad (3)$$

where c is the lattice parameter of the strained films calculated from the X-ray diffraction data and c_0 is the lattice parameter of bulk ZnO.

For *0.05M* this value is equal *+0.1*, however it is equal *-0.19* for *0.2M*, $\varepsilon_{zz}>0$ suggesting that the film is subjected to tensile stress in the plane of the substrate surface with *0.05M*, and to compressive stress with *0.2M*.

In thin films, strains originate mainly from the mismatch between the polycrystalline film and the amorphous substrate and/or from the difference in the thermal expansion coefficients of the film and the substrate [7].

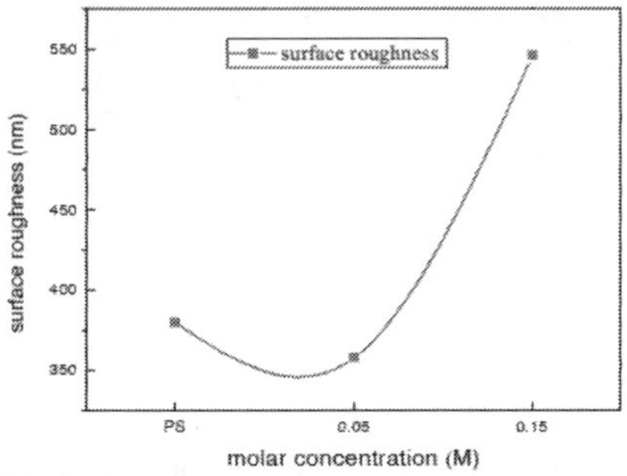

Figure 2. Surface roughness of ZnO/PS films as a function of molar concentration.

3.2 Morphogical studies

As it shown in Figure 2, the surface roughness of ZnO/PS films with *0.05M* is equal to *358nm*, however, it is equal to *546nm* when the molar concentration is *0.2M*. This result is due to the increase of size crystal, which leads to reduce the grain boundary, so the path of light stretches [8].

3.3 Optical studies

Figure 3 shows the optical reflectance spectra *R(%)* of the ZnO/PS heterojunction of two samples fabricated by different molar concentration of Zn precursors, *0.05M* and *0.2M* and the Porous silicon film without treatment. The average reflectance of the porous silicon thin film without treatment is *13%*. This value has increased to *17%* when a layer of *0.05M* is deposited, but it is returned to decrease to *10%* when layer of *0.2M* is deposited. The reduction of the reflectivity is due to roughly ordered structure originating from pores formation. Low reflectance improves the efficiency solar cell because of increase in photons absorption.

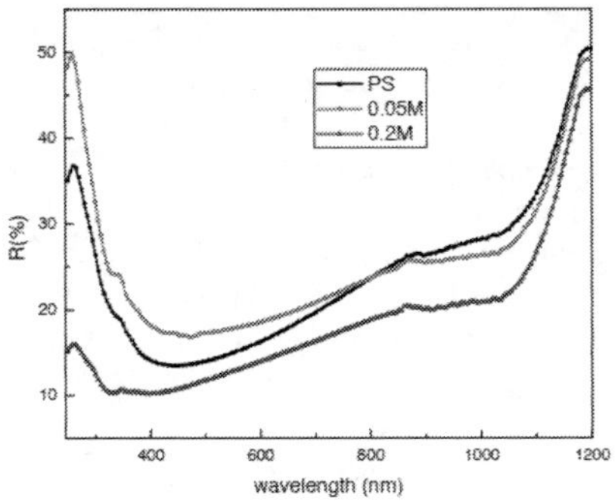

Figure 3. Total reflectivity of ZnO/PS.

3.4 Electrical studies

The I-V curve for ZnO/ PS nanoparticle thin film heterojunction device was measured. The obtained results are depicted in Figure 4. The nonlinear characteristics of the device indicate the typical diode behavior of the device to perform the current–voltage analysis of the diode. This is conforming to the equation of single junction solar cell, double-diode equivalent circuit [9]:

$$I = I_{ph} - I_{s1}\exp\left(\frac{q(V+IR_s)}{KTn} - 1\right) - I_{s2}\exp\left(\frac{q(V+IR_s)}{KTm} - 1\right) - \frac{V+IR_s}{R_{sh}} \quad (4)$$

where I_{ph} is light-generated current or photocurrent, q the electron charge $(=1,6\ .10^{-19}C)$, K is the Boltzmann's constant $(=1.38.10^{-23}J/K)$, T is the cell's operating temperature in Kelvin (K), n is the ideality factor of the first diode and m is the ideality factor of the second diode, R_{sh} is the shunt resistance, and R_S is the series resistance, The current I_{S1} and I_{S2} are called the diode saturation dark current and it depends on the reverse saturation current.

The ideality factors n and m can be calculated from the slope of the straight line region of the forward bias *ln(I)-V* plot , the equation which gives n or m can be written as follows [2] :

$$n = \frac{q}{KT} \times \frac{dV}{d\ln(I)} \quad (5)$$

Table 1 illustrates the variation of the ideality factors n and m before and after deposing ZnO layer with low and high molar concentration.

After carrying out the structural characterization of the as-grown ZnO thin film with *0.05* and *0.2* molar concentration, it can be stated that the higher molar concentratmion *0.2M* enhances structural properties of the ZnO thin film deposited on glass substrate. Through the roughness surface and reflectance studies of ZnO/PS heterojunction film, the superior optical characteristics of ZnO/PS with higher molar concentration are obvious and the electrical studies demonstrated that the efficiency of this solar cell is improved.

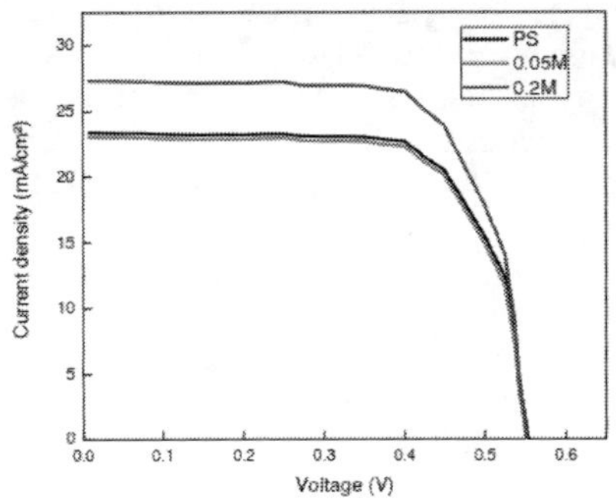

Figure 4. I-V characterization of ZnO/PS films before and after treatment.

By using equations (4) to (9) given in reference [10], in which series number of cell, *Ns=54*, cell temperature, *T=30°C*, solar irradiance *G=1000W/cm²*, series resistance, $R_s=0.221$ and shunt resistance is $R_{sh}=4150$, we have succeed to simulate I-V curves of solar cell fabricated with *0.2M* similar to those obtained in experiment, as it's shown in Figure 5.

Table1. Ideality factors of different solar cells.

Sample	Porous silicon	*0.05M*	*0.2M*
Factor ideality n	*7.63*	*6.6*	*5.5*
Factor ideality m	#	*17.7*	*20*

$$V_{OC} = \frac{KT}{q}\ln\left(\frac{I_{ph}}{I_s}\right) \quad (6)$$

$$I_{r2} = \frac{I_{ph}}{\exp\left(\frac{qV}{KT_c n} - 1\right)} \quad (7)$$

$$I_{s1} = I_{rs1}\left(\frac{T_c}{T_{cref}}\right)^3 \exp\left(\frac{qE_{g1}\left(\frac{1}{T_{cref}} - \frac{1}{T_c}\right)}{Kn}\right) \quad (8)$$

$$I_{s2} = I_{rs2}\left(\frac{T_c}{T_{cref}}\right)^{\frac{3}{2}} \exp\left(\frac{qE_{g2}\left(\frac{1}{T_{cref}} - \frac{1}{T_c}\right)}{Km}\right) \quad (9)$$

In order to understand the behaviors of ZnO/PS fabricated with *0.2M*, we have studied the impact of several parameters, by varying the solar irradiance between *500* and *1000W/m²* and the temperature between *25°C* and *35°C*. As it is shown in Figure 6, the current density is widely influenced by solar irradiance increasing. When the solar irradiance has increased from *500* to *1000W/m²*, the density current increased from *13.53mA/cm²* to *27mA/cm²*, whereas the voltage has increased from *525mV* to *546mV*, and the efficiency from *5.1%* to *10.4%*.

 https://doi.org/10.1051/matecconf/201819100013

Therefore, when the device has absorbed more photons, it can be able to generate more electron-pairs in the depletion semiconductor layer. This indicates that the light illumination increases the generation of electron–hole pairs. Illuminating the solar cell, electrons in the valence band of the ZnO absorb energy, and they are able to jump to the conduction layer.

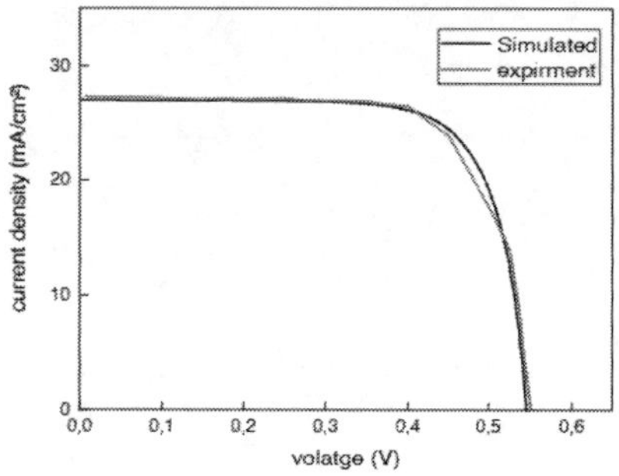

Figure 5. I-V curves of *0.2M*-solar cell.

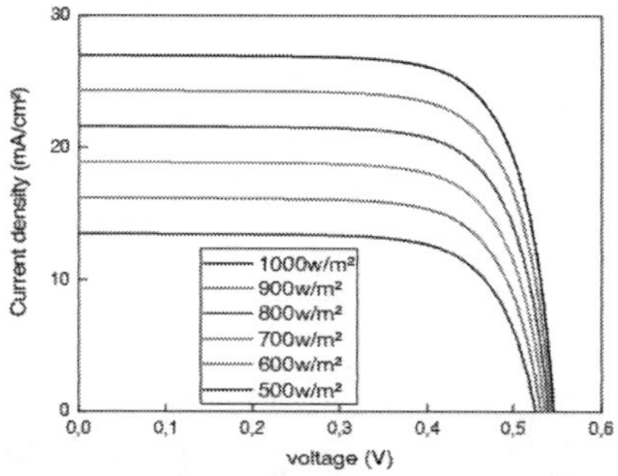

Figure 6. I-V curves of ZnO/PS under different solar irradiations

The value of voltage is influenced by varying the temperature of PV cell, however the current density is unchanged. As can be seen in Figure 7, voltage has decreased when the temperature increases. This result might be explained by the decrease in the band of semiconductor because of the increase of the energy of the electrons in the material.

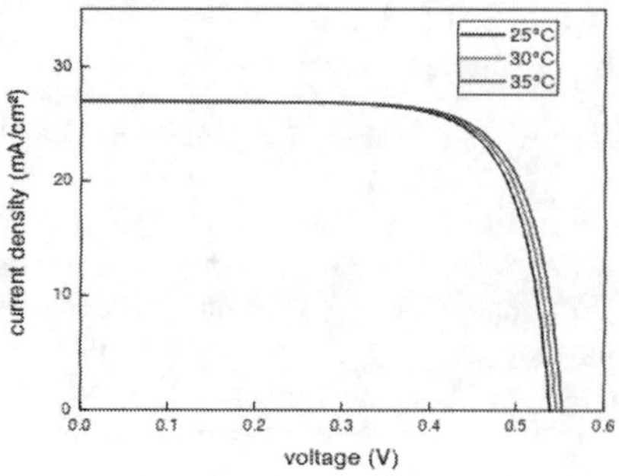

Figure 7. I-V curves of solar cell with various temperatures.

4 Conclusions

The heterojunction solar cell of ZnO/PS thin films was successfully fabricated by varying molar concentration of precursor. ZnO/PS fabricated from *0.2M* demonstrated the best proprieties structural, optical and electrical. Its surface roughness was *546nm*, reflectivity was *10%*, current density *27mA/cm²*, voltage *554mV* and its efficiency was *10.7%*. The impact of solar irradiance on this cell was studied; the current density and voltage increased largely when the solar irradiance has increased. However, when the temperature has increased the density of current was unaffected and the voltage was decreased.

References

1. W. Chebil, A. Fouzri, A. Fargi, B. Azeza, Z. Zaaboub, V. Sallet,Mater. Res. Bull., **70** (2015)
2. N. Zebbar, Y. Kheireddine, K. Mokeddem, A. Hafdallah, M. Kechouane, M.S. Aida, Mater. Sci. Semicond. Process., **14** (2011)
3. M. Salem, Z.Y. Alami, B. Bessais, A. Chahboun, M. Gaidi, , J. Nanoparticle Res., **17** (2015)
4. V. Gupta, A. Mansingh, J. Appl. Phys., **80** (1996)
5. L. Yang, B. Duponchel, R. Cousin, C. Gennequin, G. Leroy, J. Gest, J.-C. Carru, Thin Solid Film, **520** (2012)
6. A.C. Rastogi, S.B. Desu, P. Bhattacharya, R.S. Katiyar, **13** (2004)
7. R. Swapna, M. Ashok, G. Muralidharan, M.C. Santhosh Kumara, Journal of Analytical and Applied Pyrolysis, **102** (2013)
8. M.-S. Kim, B.-G. Kim, J. Kim, ACS Appl. Mater. Interfaces, **1** (2009)
9. M. Catelani, L. Ciani, M.K. Kazimierczuk, A. Reatti, Measurement, **88** (2016)
10. S.K. Cheung, N.W. Cheung, Appl. Phys. Lett., **49** (1986)

Moroccan Cedar softwood study: Application of FT-Raman spectroscopy

Somia Fellak[1,a] and Abdellatif Boukir[1]

[1] *Department of Chemistry, Faculty of Sciences and Techniques at Fez, SMBA University, Fez, Morocco*

Abstract. As non-destructive technique, FT-Raman spectroscopy has been used to study the molecular structure and monitor changes in the composition of carbohydrates and lignin components containing wood materials. For this purpose, four samples originated from Moroccan cedar wood were analyzed. Following the FT-Raman spectra, it was found that carbohydrates were identified by the bands at 898, 1098, 1123 and 1456 cm^{-1}, while lignin matrix was evaluated by the bands at 1657, 1598 and 1267 cm^{-1}. The decrease of the intensities related to these feature bands reflects the effects of natural degradation phenomenon and shows the evidence of chemical changes and quick deterioration of these contents upon exposure time to natural degradation process. Thus, the FT-Raman tool has the potential to be one of crucial sources to characterize composite materials and to evaluate the chemical changes occurred on their structures under the influence of physico-chemical or biological attacks without causing any damage of the wood surfaces or their supports.

1 Introduction

Moroccan softwood, as one of the most abundant materials on earth, is provides a resource of great value for construction and production of novel objects since antiquity. It is extensively used for many applications (artworks, packaging industry, shipbuilding, furniture, paper pulp, eating utensils, etc) since antiquity.

These materials have a heterogeneous and complex structure, primarily consisting of cellulose, lignin and hemicelluloses components which are reported as one group of materials with a well-known reputation for susceptibility to natural deterioration. The exposition to combined conditions of physical, chemical and microbial attack as ultraviolet (UV) light, solar irradiation, moisture (humidity), temperature and fungus can cause molecular degradation of their main components. It results a loss of fiber strength and rigidity. This is manifested in lower mechanical stability that may leads, sometimes, to full disintegration of wooden materials [1], and consequently results in loss of cultural heritage [2].

Hence, an accurate characterization and examination of different changes occurred in these materials through a spectroscopic study by non-destructive techniques, is extremely important for optimal safeguarding and preservation.

A very few works have been conducted on structural characterization and/or structural degradation of wood by Fourier transform Raman spectroscopy (FT-Raman) [3-6], in contrast to the other spectroscopic techniques which are the most used for material's study as Fourier-transform infrared spectroscopy (FTIR), X-ray diffraction (XRD), scanning electron microscopy (SEM) and nuclear magnetic resonance (NMR). In this field, Agarwal and Ralph [3] have applied FT-Raman technique to identify the major constituents of black spruce wood: lignin, cellulose and hemicelluloses, while Ona et al. [7] have performed interesting researches on the Eucalyptus wood properties by the same technique.

Recently, FT-Raman is being more performed for the chemical analysis of biomaterials as wood [8-9]. It provides more information about polymer chain and fundamental knowledge at a molecular (micro-level) and macro-level [10]. In addition, it has been shown to be a valuable technique for analyzing structural changes in the fibers which arise from physical, chemical or mechanical processing [11-12]. Hence it is possible to perform fast, non destructive and non invasive measurements without extensive sample preparation [9].

As this non-destructive spectroscopic method seems to be a promising instrument for studying composition of wood materials, the main goal of the present work is to investigate, in details, the distribution of chemical composition in softwood materials and understand the structural rearrangement caused by the effect of natural degradation process.

[a] Corresponding author: soumaya.fellak@gmail.com

1 Materials and methods

1.1 Sampling

Analyzed Softwood samples were collected from four archaeological Cedar wood (*Cedrus atlantica*) dating to the 18th, 19th, 20th and 21st centuries in the Ecomuseum of Tazekka under standard climate. The pieces originate from Tazekka national park (WGS84: 34°6′0″N, 4°11′0″W) located in the Middle Atlas of Morocco and near the city of Taza (Bab Boudir region). The samples were dated by specialist researchers using Radiocarbon dating method. Thus, the dimensions of wood samples are 200×200×100 mm^3 (tangential × radial × longitudinal directions). The characteristics of the experimental materials are presented in Table 1. The FT-Raman analysis was performed directly on the surface of the samples.

Table 1. Sample's description.

Sample	**Age** (century)
D_1	21th
D_2	20th
D_3	19th
D_4	18th

1.2 FT-Raman spectroscopy

The FT-Raman study was conducted with a Bruker (USA) MultiRAM Stand Alone FT-Raman Spectrometer. The instrument is equipped with a diode-pumped Nd:YAG excitation source with a large emission intensity at 1.064 nm. Furthermore, the signal was collected with a liquid nitrogen cooled germanium detector. For each FT-Raman measurement 100 scans were averaged, with a resolution of 4 cm^{-1} and a time measurement of 3 min for each spectrum. All FT-Raman spectra were registered from 4000 to 250 cm^{-1}. Three analyses were performed on several locations for each sample. The temperature and humidity room were controlled during analysis.

3 Experimental results

The common bands assignments of four wood's sample (D_1, D_2, D_3 and D_4), are given in Table 2. The band's attribution was extremely difficult due to the overlapping of some cellulose and lignin bands; and so, it confirmation was based on different literature data [5, 6, 13] which focused on the wood study and investigation of degradation effect using FT-Raman spectroscopy.

Table 2. Assignment of characteristic FT-Raman bands of softwood materials (D_1, D_2, D_3 and D_4) [13-21].

Wavenumber (cm^{-1})	Assignment
2943	□ CH (asymmetric) in OCH_3 of lignin
2895	CH_2 group in the glucopyranose ring of cellulose I_β
1657	Conjugated C=C stretching of coniferyl alcohol (guaiacyl) in lignin that overlaps with C=O stretch of coniferyl aldehyde
1598	□ C=C aromatic skeletal vibration typical for lignin guaiacyl and syringyl monomer in lignin
1456	HCH bending and small proportion of HOC bending in pure or amorphous cellulose
1378	δ(C-H) and δ(CH_2) in cellulose and hemicelluloses
1333	C-H vibration in cellulose and C-O vibration in syringyl derivatives
1267	C_{ar}-O of aryl-O-CH_3 and/or aryl-OH of guaiacyl/syringyl monomer aromatic ring
1123	Stretching vibration of C-O, C-O-C glycosidic linkage in cellulose and xylan or C-C ring
1092	
898	CH deformation in amorphous cellulose
379	δ_s(CCC) in crystalline cellulose

Figure 1 reports the representative FT-Raman spectra for each sample (D_1, D_2, D_3 and D_4) between the spectral region of 3500-500 cm^{-1}.

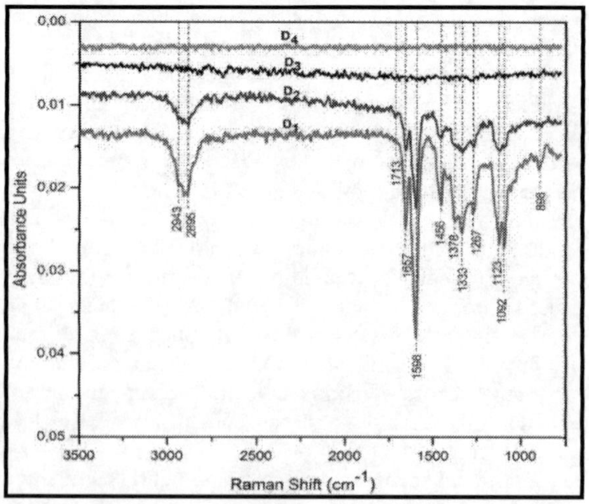

Figure 1. FT-Raman spectra acquired from the four samples of softwood: D_1- Wood sample dating to 21st century; D_2- Wood sample dating to 20th century; D_3- Wood sample dating to 19th century; D_4- Wood sample dating to 18th century.

3.1 Cellulose and hemicelluloses

The main characteristic FT-Raman signals for cellulosic and hemicellulosic fibers can be visually divided into

 https://doi.org/10.1051/matecconf/201819100014

two essential regions: 3500-2800 cm^{-1} and 1480-250 cm^{-1} fingerprint region (Fig. 1 and 2).

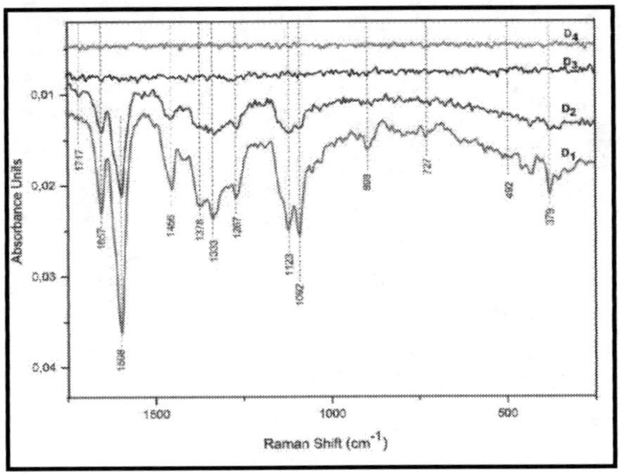

Figure 2. FT-Raman spectra 1750-250 cm^{-1} range acquired from the four samples of softwood: D_1- Wood sample dating to 21st century; D_2- Wood sample dating to 20th century; D_3- Wood sample dating to 19th century; D_4- Wood sample dating to 18th century.

For the first region, the detected bands are mainly due to the hydroxyl groups, methyl and methylene stretching vibrations. Concerning the second range, the bands correspond to methylene, methyl bending, wagging, rocking, C-O-H in-plane bending and C-O-H in-plane bending and skeletal bending vibrations (CCC, COC, OCC and OCO). Thus, the deterioration of the cellulose and hemicelluloses fractions has been explained by the decline in intensities of these bands (Fig. 1) during exposure to natural atmospheric effect.

The recent samples dating to 21st and 20th century (Fig. 1. D_1 and D_2) clearly display a feature band at 2895 cm^{-1}. According to Barnette et al. [13], the latter was attributed to symmetric stretching vibrations of the CH_2 group in the glucopyranose ring of cellulose I_β. The presence of this band in spectra of D_3 and D_4 (Fig. 2) degraded samples dating to 19th and 18th century respectively, suggests that upon a long exposure to the degradation phenomenon, the crystalline fraction decomposes and results disorder fraction which in turn re-crystallized and formed the new ordered fraction. This finding can be confirmed by the decline in intensities of feature bands typical of amorphous cellulose at 1456 cm^{-1} and 898 cm^{-1} and assigned to HCH bending and small proportion of HOC bending in amorphous cellulose as well as CH deformation in amorphous cellulose, respectively [14]. On the other hand, the C-H and CH_2 deformations in cellulose and hemicelluloses compounds were observed at 1378 cm^{-1}.

From spectra of sample D_1 and D_2 (1200-1000 cm^{-1}), it is easily to distinguish a doublet of peaks at 1123 and 1092 cm^{-1} assigned to combined stretching vibration of C-O ring and C-O-C glycosidic linkages in cellulose and hemicellulose [13, 15]; providing information about the breaking of cellulosic chains at the β-1,4-glycosidic ether bonds. Thus, the disappearance of these two bands in spectra of oldest samples (Fig. 2 D_3 and D_4) indicates the serious degradation occurred on cellulose and hemicelluloses.

Referring to literature data [16], the bands at 379 cm^{-1} can be unambiguously attributed to CCC deformations in crystalline fraction of cellulose. Fig. 2 shows a discernible decrease of this band proportionally with the age of sample, indicating the loss in mechanical rigidity and toughness for these materials, consequently, the surrender of lignocellulosic biomass against deconstructive processes.

The weak features at 379 and 440 cm^{-1} might be explained as a result of intermolecular interactions between lignin and carbohydrates, that can caused a small shifts in peak positions and/or changes in band shapes [3].

3.2 Lignin

In order to estimate lignin fraction, different bands were studied. In the region between 3100-2800 cm^{-1}, the band at 2943 cm^{-1} was attributed to the C-H stretching of the methoxy groups in lignin [5, 6]. It appears less pronounced in the oldest samples dating to the 19th and 18th centuries (Fig. 1 D_3 and D_4) indicating lower lignin presence compared to the youngest ones (Fig. 1 D_1 and D_2).

The detection of feature band at 1717 cm^{-1} in spectra of samples D_2, D_3 and D_4 (Fig. 2) indicates the presence of carbonyl groups related to the residual lignin amount resulted from delignification of wood sample upon process of natural degradation. It relative intensity appears no changeable for all oldest samples, while it appears absent in D_1 spectrum (Fig. 2). The non important sensitivity of this produced amount to degradation events is the possible explanation in this case.

Furthermore, the combined band in the region between 1657 and 1598 cm^{-1} was mainly originates from guaiacyl (coniferyl alcohol units for softwood) and syringyl (sinapyl alcohols units for hardwood) matrix in lignin compound. The band detected at 1657 cm□-□1 is attributed to conjugated C=C stretching vibration of coniferyl alcohol (guaiacyl) in lignin that overlaps with C=O stretch of coniferyl acid after oxidation of alcohol in side chain [6, 17]. According to Kihara et al. [18], this band can also assigned to marker bands for conjugate carbonyl groups (α,β-unsaturated C=O).

The most intense peak at 1598 cm^{-1} (Fig. 2 D_1 and D_2) is attributed to stretching vibration of polar aromatic C=C in phenolic compounds [19] related to guaiacyl and syringyl monomers in lignin [5, 20].

The other predominant lignin band was detected at 1267 cm^{-1} and corresponds to $C_{aromatic}$-O of guaiacyl lignin for softwood [17]. It shifted to lower intensities during exposure time to natural degradation process (Fig. 2), because of guaiacyl lignin is less susceptible than syringyl lignin. Nevertheless, for hardwood spectra, there is a rapid decline in its intensity. Thus, we can confirm that our cedar samples belong to the softwood specie.

It is likely reported that decomposition of hemicelluloses and/or extractives, can lead to the decrease in quantity of lignin, consequently, a simultaneous deterioration of wooden materials.

4 Conclusions

The present work has put in evidence the crucial role of FT-Raman spectroscopy as non destructive method to characterize and study the effect of natural degradation on chemical structure of cellulose, hemicelluloses and lignin as major components of softwood by providing accurate information about their chemical structures. Based on the obtained results, the gradual decline in intensities of features bands related to these constituents for each compound of wood, suggest their sensitivity to combined degradation agents and, consequently, irreversible losses of softwood material in archaeological sites.

The authors gratefully acknowledge technical support by National Center for Scientific and Technical Research in Morocco. We thank the Ministry of Culture, Arts and Museum Division for sampling step.

References

1. E. Marengo, E. Robotti, M. C. Liparota, M. C. Gennaro, Talanta **63** (2004)
2. T. Rosado, M. Silva, C. Pereira, J. Mirão, A. Candeias, A. Teresa Caldeira, I. J. C. S. **6** (2015)
3. U. P. Agarwal, S.A. Ralph, Appl. Spectrosc. **51** (1997)
4. Y. Xia, T.Y. Chen, J.L. Wen, Y. Zhao, J. Qiu, R.C. Sun, Int. J. Biol. Macromol. **109** (2018)
5. Z. Ji, J.F. Ma, Z.H. Zhang, F. Xu, R.C. Sun, Ind. Crop. Prod. **47** (2013)
6. Ö. Ozgens, S. Durmaz, I.H. Boyaci, H. Eksi-Kocak, Spectrochim. Acta A. **171** (2017)
7. T. Ona, T. Sonoda, K. Ito, M. Shibata, T. Kato, Y. Ootake, Y. Tamal, Y. Kojima, J. Pulp Pap. Sci. **26** (2000)
8. M.R. Almeida, C.H.V Fidelis, L.E.S. Barata, R.J. Poppi, Talanta **117** (2013)
9. V. Emmanuel, B. Odile, R. Céline, Spectrochim. Acta Part A Mol. Biomol. Spectrosc. **136** (2015)
10. U.P. Agarwal, *An Overview of Raman Spectroscopy as Applied to Lignocellulosic Materials* (Advances in Lignocellulosics Characterization, Paper Industry Management Association, TAPPI Press, Atlanta, USA, 1999)
11. H. Edwards, D. Farwell, D. Webster, Spectrochim. Acta A. **53** (1997)
12. A.P.P. Alves, L.P.Z. de Oliveira, A.A.N. Castro, R. Neumann, L.F.C. de Oliveira, H.G.M. Edwards, A.C. Sant'Ana, Vib. Spectrosc. **86** (2016)
13. A.L. Barnette, C. Lee, L.C. Bradley, E.P. Schreiner, Y.B. Park, H. Shin, D.J. Cosgrove, S. Park, S.H. Kim, Carbohyd. Polym. **89** (2012)
14. S.A. Centeno, A.Vila, L. Barro, Microchem. J. **114** (2014)
15. S. Yamauchi, Y. Iijima, S. Doi, J. Wood. Sci. **51** (2005)
16. M. Zhang, C. Lapierre, N.L. Nouxman, M.K. Nieuwoudt, B.G. Smith, R.R. Chavan, B.H. McArdle, P.J. Harris, Plant. Physiol. Bioch. **118** (2017)
17. A. Cogulet, P. Blanchet, V. Landry, J. Photochem Photobiol B. **158** (2016)
18. M. Kihara, M. Takayama, H. Wariishi, H. Tanaka, Spectrochim. Acta A. **58** (2002)
19. A. Boukir, M. Guiliano, L. Asia, G. Mille, Analusis. **26** (1998)
20. S. Yamauchi, Y. Tamura, Y. Kurimoto, A. Koizumi, J. Adhes. Soc. Jpn. **33** (1997)
21. B. Zghari, P. Doumenq, A. Roman, A. Boukir, J. Mater. Environ. Sci. **8** (2017)

AUTHOR INDEX